Y0-BVO-322

The Palgrave Environmental Reader

The Palgrave Environmental Reader

Edited by

Daniel G. Payne

and

Richard S. Newman

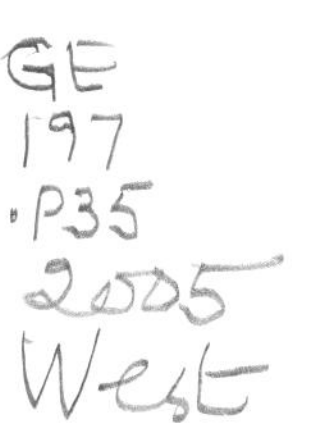

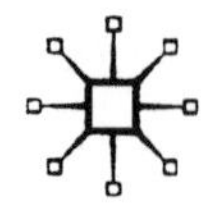

THE PALGRAVE ENVIRONMENTAL READER

© Daniel G. Payne and Richard S. Newman, 2005.

All rights reserved. No part of this book may be used or reproduced in any manner whatsoever without written permission except in the case of brief quotations embodied in critical articles or reviews.

First published in 2005 by
PALGRAVE MACMILLAN™
175 Fifth Avenue, New York, N.Y. 10010 and
Houndmills, Basingstoke, Hampshire, England RG21 6XS
Companies and representatives throughout the world.

PALGRAVE MACMILLAN is the global academic imprint of the Palgrave Macmillan division of St. Martin's Press, LLC and of Palgrave Macmillan Ltd. Macmillan® is a registered trademark in the United States, United Kingdom and other countries. Palgrave is a registered trademark in the European Union and other countries.

ISBN 1–4039–6593–5
ISBN 1–4039–6594–3

Library of Congress Cataloging-in-Publication Data

The Palgrave environmental reader / edited by Richard S. Newman and Daniel G. Payne.
p. cm.
Includes bibliographical references and index.
ISBN 1–4039–6593–5—ISBN 1–4039–6594–3
1. Environmentalism—United States. I. Newman, Richard S. II. Payne, Daniel G., 1958–

GE197.P35 2005
333.72—dc22 2005048666

A catalogue record for this book is available from the British Library.

Design by Newgen Imaging Systems (P) Ltd., Chennai, India.

First edition: November 2005

10 9 8 7 6 5 4 3 2 1

Printed in the United States of America.

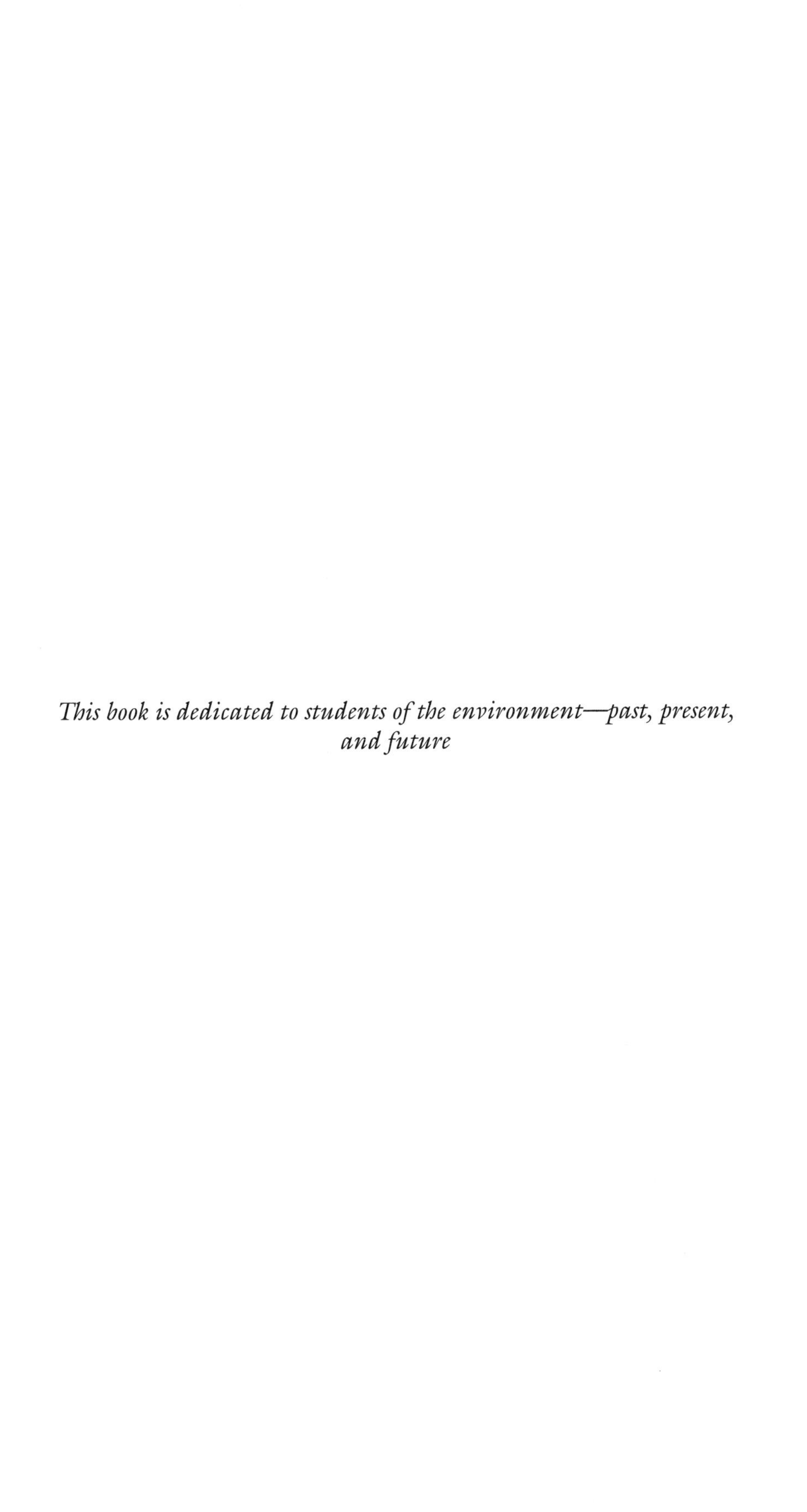

This book is dedicated to students of the environment—past, present, and future

Whatever attitude to human existence you fashion for yourself, know that it is valid only if it be the shadow of an attitude to Nature. A human life, so often likened to a spectacle upon a stage, is more justly a ritual. The ancient values of dignity, beauty, and poetry which sustain it are of Nature's inspiration; they are born of the mystery and beauty of the world. Do no dishonour to the earth lest you dishonour the spirit of man.

Henry Beston
The Outermost House

Contents

Acknowledgments

The editors wish to express our appreciation to Brendan O'Malley for believing in the project from the start. Brendan's sound editorial suggestions at an early stage were enormously helpful to us throughout the process of putting this work together. We also wish to thank the fine editorial staff at Palgrave Macmillan for their expert help. Thanks also to the students at SUNY Buffalo, SUNY Oneonta, and RIT who provided valuable feedback regarding the selection of some of these documents and to Professors Robert Newman and James Bunn of SUNY Buffalo's English Department for allowing both authors to test material in the book. Grant funding to defray the cost of copyright expenditures was provided by Andrew Moore, Dean of the College of Liberal Arts at the Rochester Institute of Technology; the Walter B. Ford Grants Committee at SUNY College at Oneonta; and the Professional Development Committee of the SUNY Oneonta chapter of United University Professions.

Sincere thanks also goes to the teachers, mentors, and friends who have provided such valuable guidance to me along the way: Clinton Trowbridge (Dowling College); Jim McCord (Union College); Marcus Klein and Robert Newman (SUNY Buffalo); John Elder (Middlebury College); and John Tallmadge (Union Institute). Heartfelt appreciation also goes to my friends and colleagues at SUNY Oneonta, the Association for the Study of Literature and the Environment (ASLE), and the John Burroughs Association, as well as to my wonderful (and extraordinarily tolerant) family.

D.G.P.

Many thanks to Luella Kenny for her insights and comments on fieldtrips to Love Canal. And to RIT's College of Liberal Arts for funding to research some of these documents. Thank you, too, Lisa for coming into my life.

R.S.N.

Introduction

On June 22, 1969, an oil slick and other debris wedged under a railroad trestle on a portion of the Cuyahoga River in downtown Cleveland caught fire. The incident soon became a vivid symbol of how badly polluted many of America's waterways had become; less than one year later, more than 22 million Americans celebrated the nation's first Earth Day. Responding to the outcry by a public that was becoming increasingly alarmed by widespread environmental problems, Congress created the Environmental Protection Agency (EPA) in 1970 and passed a series of sweeping environmental reforms, including the National Environmental Policy Act (1969), the Clean Air Act (1970), Occupational Safety and Health Act (1970) the Environmental Pesticide Control Act (1972), the Toxic Substances Control Act (1976), and the Clean Water Act (1977). Perhaps the most remarkable law arising out of this era of environmental reform was the Endangered Species Act of 1973, where, for the first time, legislative protection was extended to include species and ecosystems whose preservation would have little or no direct benefit to humans. The impetus for environmental reform continued unabated throughout the decade, spurred on by such incidents as the Love Canal disaster, which led to the establishment of the Comprehensive Environmental Response, Compensation and Liability Act of 1980 (CERCLA), more commonly known as Superfund, which was formed to clean up and determine liability for hazardous waste sites.

The environment continues to draw attention as concern spreads over issues such as global warming, environmental justice, biodiversity, and the depletion of the ozone layer. Open the newspaper, surf on-line, open your mail—news about the environment abounds: "Plan Gives Farmers a Role in Fighting Global Warming," reads one recent headline in *The New York Times*, alongside another story on how "South Florida Freezes are Linked to Draining Wetlands."[1] On "Marketplace," a public-radio program about the American economy, many environmental stories have appeared in recent months, like the one on November 3, 2003, about Fund managers' worries about global warming. Are the companies they invest in secure from environmental lawsuits? The extent to which concern about the environment pervades our national discourse may best be epitomized by the box office success of the 2004 global warming disaster film *The Day After Tomorrow*—and the extent to which the reelection campaign of President George W. Bush was forced to respond to the film's scenario of environmental catastrophe caused by the greenhouse effect. Even at the most local level, environmental matters have taken a prominent place: Pesticide certification laws require homeowners or lawn care companies to inform neighbors of their spraying plans; suburban and exurban development fosters public debate over sprawl and spoiled scenery; the use of former industrial sites (brownfields) raises the issue of restitution for harm done to citizens' health. The environment, it seems clear, is not just for environmentalists any more.

While the past thirty years have seen a remarkable growth of public and scholarly interest in the environment, a close examination of our shared history shows that the environment (in the largest sense of the word) has long been one of the defining features of American political discourse. The astonishing natural resources of the North American continent were remarked on by numerous writers of the early colonial period, such as John Smith, Thomas Harriot, Francis Higginson, and William Wood, to name just a few. As John Smith wrote in *A Description of New England* (1616), "What pleasure can be more than [for potential settlers] to recreate themselves before their own doors in their own boats upon the sea, where man, woman, and child, with a small hook and line, by angling, may take divers sorts of excellent fish at their pleasure? And is it not pretty sport to pull up two pence, six pence, and twelve pence, as fast as you can haul and veer a line?"[2] Less than twenty years after the landing at Plymouth, however, the Massachusetts Bay Colony in Boston was already suffering from a local shortage of firewood, with John Winthrop describing in a 1638 journal entry how a wood gathering expedition to Spectacle Island had resulted in one fatality and several cases of frostbite. Go back to newspapers, diaries, and letters from bygone eras and you see environmental issues quite prominently displayed. In Ben Franklin's Pennsylvania Gazette of the 1740s, the civic-minded inventor told his readers about the many advantages of his newly invented "Pennsylvania Fireplace." With wood supplies dwindling in northern colonies, he noted, a premium would be placed on any technology that burned fuel more efficiently and cheaply.

Cases such as these were, however, primarily local and isolated. The predominant ideology of colonists and early Americans valorized the conquering of the land. In response to criticism by European observers such as Peter Kalm regarding the colonists' wasteful agricultural practices, Thomas Jefferson pointed out that the exigencies of colonial life made such practices not only understandable but necessary: "The indifferent state of [agriculture] among us does not proceed from a want of knowledge merely; it is from our having such quantities of land to waste as we please. In Europe the object is to make the most of their land, labour being abundant; here it is to make the most of our labour, land being abundant."[3] Such defenses of the "wasteful" practices of American settlers have a degree of truth to them; still, it is hard to dispute Alexis De Tocqueville's assessment of the issue in *Democracy in America* (1830): "In Europe people talk a great deal about the wilds of America, but the Americans themselves are insensible to the wonders of inanimate nature, and they may be said not to perceive the mighty forests around them till they fall beneath the hatchet. Their eyes are fixed upon another sight: the American people views its own march across these wilds,—drying swamps, turning the course of rivers, peopling solitudes, and subduing nature."[4]

Despite such criticism, few eighteenth- and nineteenth-century Americans questioned the notion that "subduing nature" was desirable. This mindset, of course, differed sharply from that of Native Americans. Precolumbian America contained a host of native cultures, each with their own sense of the landscape, but sharing numerous ecological perspectives. While modern environmental historians such as William Cronon have convincingly argued that Native American environmental practices also changed the land, they generally tended to adapt agricultural and settlement practices to conform to the landscape rather than change it. From the Northeast Woodlands to the Southwest, native peoples viewed themselves as an integral part of the landscape. This ideology bears a striking resemblance to modern ecological perspectives such as that of Deep Ecology, and indeed, native cultures influenced many modern-day environmental writers. While there is a dearth of direct Native American testimony from early America, one of the most powerful and widely circulated comments on the centrality of land in Native American life comes from the Seneca orator and diplomat Red Jacket in 1805 who invoked both practical and spiritual arguments in his counsel against further land dealings with Americans.

By the mid-1800s, deforestation in the east and the remarkably rapid transformation of the frontier wilderness into farms and cities prompted writers such as George Catlin, Henry Thoreau, George Perkins Marsh, William Bryant, and John Muir to call for the establishment of national parks and wildlife preserves. As early as the 1830s, William Cullen Bryant had observed that Americans might not be able to enjoy the great scenes that many took for granted. Governments could—and must—act by setting aside space for national parks and forest preserves. These early calls went largely unheeded—indeed, the pressure for expansion and development was greater than ever before. Technology was often at the center of the debate. Whereas prior to 1800, the "environment" signified the landscape beyond humans (whether local or global, philosophical or economic), by the middle of that century the "environment" could mean man-made structures like canals, railroads, bridges, and factories. If the term environment was not yet in common usage, Americans nevertheless spoke about (and wrote about) these artificial creations in environmental terms. As historian Carol Sheriff points out, America lexicon about the environment had shifted already by the 1820s and 1830s with the completion of the Erie Canal. Writers referred to the canal not merely in traditional nomenclature terms ("Clinton's Ditch," "A waterway") but in ultramodern ones. It was the "artificial river," a 363-mile long man-made canal. Connecting the Hudson River system with the Great Lakes, the canal was much more than nature extended. Rather, it was nature transcended. If nature had not put an economically beneficial east-west river system in New York, then humans would do so. The impact was swift and wide-ranging, for New Yorkers and Americans began to re-orient their lives around this artificial canal: farmers moved as close as possible to it and produced more for the market; cities sprouted in areas formerly viewed as inhospitable for their inland location. The very concept of time and distance was shattered, as long journeys were reduced to days and long distances reduced to mere time on the canal. Not everyone was enthusiastic about the change. Novelist Nathaniel Hawthorne, for one, complained that the artificial canal despoiled the surrounding natural environment, as swamps and tributaries were drained to fill the Erie Canal. Hawthorne contrasted the artificial vibrancy of the canal with the desolate, cemetery-like look of the trees and grasslands around it, but, as Sheriff concludes, Hawthorne was decidedly in the minority.[5] More Americans wanted canals and then railroads and then highways, cars, airplanes. By the 1870s, railroads had brought more people to Niagara Falls in a single year than had visited it in decades before the Civil War. Thus, the Falls could be seen by millions of Americans outside of Western New York (or Canada) as a national, or even global, treasure to be saved from hucksters and developers.

Such concerns coalesced in the late nineteenth century with the growth of the early conservation movement, which culminated in the Progressive era policies of Theodore Roosevelt. While there were significant policy disputes during this period between the "wise-use" policies of conservationists such as Roosevelt and his chief forester Gifford Pinchot and more ardent advocates of wilderness preservation such as John Muir, both factions of this "first generation" of environmental reform were concerned primarily with land use issues such as resource conservation, wilderness preservation, and the establishment of national parks and monuments. It was also during the Progressive era, however, that some of the more complex "second generation" environmental issues begin to surface. In 1905 Upton Sinclair examined the unsanitary practices of the meatpacking industry in his novel *The Jungle*, leading to the passage of the Pure Food and Drugs Act of 1906. In the early 1900s, a young engineer named Albert Hooker argued that there was an ecological benefit to his family's innovations in chemical technology. That technology of hydro-chemicals would later become associated with Love Canal—but in his 1914 text, Hooker declared that chemicals could clean waters fouled by human and industrial waste.

Starting with the dawn of the atomic era in 1945, second-generation environmental issues, such as nuclear war, disposal of nuclear wastes, chemical contamination, global warming, biodiversity and so on, have superceded land use policies as the primary focus of many environmentalists. The publication of Rachel Carson's *Silent Spring* in 1962 had a galvanizing effect in calling public attention to the potential hazards of technologies that had hitherto been seen as benign, even beneficial, to human beings and their environment. The backdrop of any modern environmental text is surely the remarkable rise of environmental movements in the latter half of the twentieth century. If environmental concerns were present up to the middle of the twentieth century, they have since intensified to the point where they are consistently cited in public opinion polls as one of the primary concerns of the American public. The impact of this interest has been far-reaching.

Over the past thirty years in particular, environmental laws have been passed at an impressive rate. In addition, the number of environmental groups has grown considerably. Finally, and perhaps most important for our purposes, environmental studies has emerged as a fully developed field of study encompassing numerous disciplines, including history, science, literature, philosophy, economics, and law.

With all of this new work in environmental studies and law, we might ask how Americans' understanding of their environment has changed over time? One trend has been the movement from localism to global environmental concerns and back—to a certain extent, at least—localism once again. Looking at the history of air pollution, for instance, environmental historian J.R. McNeill has written that "For most of history [the issue] had existed only locally and had had only modest consequences."[6] After 1900, however, "its scale grew exponentially." By the 1970s, air pollution was an issue for not just localities (Los Angeles or Pittsburgh, say) but entire nations and even the global community as a whole. Increasing numbers of environmentalists have adopted the mantra of "think globally, act locally" and it may fairly be said that for many, activism begins at home; literally in the sense of the products and services people buy and in an increased emphasis on local and regional environmental issues; and metaphorically in the sense that issues such as global warming may affect our common home, the earth.

The shift from local to global concerns—and then back again to localism—is but one key theme in American environmental writing. The other remarkable point about American environmental writing is that it has been linked to democratic outreach and public policy. From the earliest environmental writers onward, the notion of the "public" has been a central concern. This makes perfect sense in a democratic republic that revolves around citizen participation. Ben Franklin's observations on the danger of lead paint offer one such starting point to this tradition. Franklin called lead an evil that must be watched. But by whom? In colonial America, where centralized governmental institutions and the Church had much less power than in Europe, Franklin turned toward the public at large as the "environmental watchdog." Given this emphasis on the democratic process, it is not surprising how important nature writers have been to the cause of marshalling support for environmental reform. The names of writers such as John Muir, Aldo Leopold, and Rachel Carson spring readily to mind but other, less politically active writers such as John Burroughs played a significant, if sometimes indirect, role in marshalling public support for environmental reform.

The environmental justice movement of the 1970s and 1980s saw new groups emerge in environmental politics, but they also often employed this strategy of reaching out to the public. Very often, as in the case of the Love Canal activists, the public was viewed as the only way to wage a battle for environmental justice, for the people who sought to present their stories were economically and politically marginalized. Modern corporations' power and political muscle, they argued, made traditional lobbying difficult if not impossible. "The best weapon you have," Lois Gibbs still tells local activists, "is your story." In this manner, Love Canal stands as a modern-day equivalent to Henry Thoreau, for the citizen activists of this ill-fated community helped redirect environmentalism. One of the earliest statements from Love Canal citizens—the Love Canal Homeowners Association statement—expressed an abiding interest in creating a public and governmental awakening to chemical depletion of the environment. "We are a group of concerned citizens and environmentalists," the group observed in August 1978, "worried about the effects of toxic wastes in our area, and across the country. Love Canal families have seen first-hand what low-level chemical exposure can do to our health and the environment." While homeowners wanted to be evacuated from Love Canal, they also sought "to educate and assist other communities with their problems . . . in the hazardous waste issue." With the EPA estimating that there were over "30,000 . . . toxic waste dumps across the country," Love Canal activists hoped to spur local governmental action and citizens mobilizations. Americans could renew their environment, the group concluded, "if you, the taxpayer, votes, citizens, for the government to use their authority! You have to put a stop to the industrial poisoning of America. Write your representative, join an environmental group in your area, contact us. Together we can fight pollution; alone we will suffer from it."

The goal of the editors of this anthology has been to explore the rich, diverse, and constantly evolving dynamic between Americans and their environment by reprinting key primary documents. In choosing the works represented in this anthology, we have attempted, wherever possible, to offer longer, more inclusive selections, opting to sacrifice breadth for depth. Second, we have included coverage of areas that go beyond nature writing by including pieces representing work in grassroots activism, environmental justice, environmental law, and science. While our primary focus is on American writers, we have also included selections from three European writers—Blackstone, Kalm, and Naess—as a means of providing important context for our other selections.

NOTES

1. David Barboza, "Plan Gives Farmers a Role in Fighting Global Warming," *New York Times*, November 25, 2003, Sec. F, Page 2. Anahad O' Connor, "South Florida Freezes are Linked to Draining of Wetlands," *New York Times*, November 25, 2003, Sec. F, Page 2.
2. John Smith, *A Description of New England*. 1616. *In Library of Puritan Writings*, vol. 9, Sacvan Bercovitch, ed. (New York, AMS Press, 1986), 126.
3. Thomas Jefferson, "Notes on the State of Virginia." 1787. *The Portable Thomas Jefferson*, Merrill D. Peterson, ed. (New York: The Viking Press, 1977), 125–126.
4. Alexis de Tocqueville, *Democracy in America*. 1835 (New York: Mentor Books, 1956), 181.
5. Carol Sheriff, *The Artificial River: The Erie Canal and the Paradox of Progress, 1817–1862* (New York: Hill and Wang, 1966), 57–58.
6. J.R. McNeill, *Something New Under the Sun: An Environmental History of the Twentieth-Century World* (New York: W.W. Norton and Company, 2000), 51.

1

William Blackstone (1723–1780)

Sir William Blackstone, an eminent jurist and professor of law at Oxford University, was the eighteenth century's leading authority on the development of English common law. The underlying principles of the common law were based primarily on the customs and usages of the society, encompassing the ancient, often unwritten, laws of England, statutory enactments, and the judgments and decrees of the English courts. Blackstone's monumental four-volume treatise on English Law, Commentaries on the Laws of England *was first published in 1765 and was soon recognized as the authoritative guide to English common law. In the thinly settled American colonies, the need for a portable yet definitive guide to the common law made Blackstone's* Commentaries *an essential treatise for colonial lawyers and magistrates. Even after the revolution, Blackstone's work remained a cornerstone of early American jurisprudence, as adapted and edited by the federalist law professor St. George Tucker in 1803.*

By the seventeenth century, English law included provisions that are in some ways recognizable as precursors to American conservation laws of the late nineteenth century, however, English forest and game laws of this period were rooted in the same soil as the feudal system, and were linked more closely to the property rights of the crown and nobility rather than to anything resembling a true conservation ethic. As Blackstone writes in his Commentaries, *"with us in England . . . hunting has ever been esteemed a most princely diversion and exercise." Given the realities of colonial life, however, any common law restrictions on the taking of game and forest use were often disregarded, and Blackstone's* Commentaries *provided support for this position as well, stating that in the absence of restrictive laws, "every man, from the prince to the peasant, has an equal right of pursuing, and taking to his own use, all such creatures as are* ferae naturae, *and therefore the property of nobody, but liable to be seized by the first occupant." Colonial legislatures were reluctant to enact or enforce measures limiting the "right" to use common resources without restriction, as English-style forest and game laws tended to be associated with elitism or class discrimination—politically volatile charges that are sometimes used even today by opponents of environmental reform.*

From *Commentaries on the Laws of England* Book II, Chapter XXVII

There still remains another species of prerogative property, founded upon a very different principle from any that have been mentioned before; the property of such animals *ferae naturae*, as are known by the denomination of *game*, with the right of pursuing, taking, and destroying them: which is vested in the king alone, and from him derived to such of his subjects as have received the grants of a chase, a park, a free warren, or free fishery. This may lead us into an inquiry concerning the original of these franchises, or royalties, on which we touched a little in a former chapter: the right itself being an incorporeal hereditament, though the fruits and profits of it are of a personal nature.

In the first place then we have already shewn, and indeed it cannot be denied, that by the law of nature every man from the prince to the peasant, has an equal right of pursuing, and taking to his own use, all such creatures as are *ferae naturae*, and therefore the property of nobody, but liable to be seized by the first occupant. And so it was held by the imperial law, even so late as Justinian's time: *ferae igitur bestiae, et volucres, et "omnia animalia quae marie, cœlo, et terra nascuntur, simul atque ab aliquo capta fuerint, jure gentium statim illius esse incipiunt. Quod enim nullius est, id naturali ratione occcupanti conceditur1.*"[1] But it follows from the very end and constitution of society, that this natural right, as well as many others belonging to man as an individual, may be restrained by positive laws enacted for reasons of state, or for the supposed benefit of the community. This restriction may be either with respect to the *place* in which this right, may, or may not, be exercised; with respect to the *animals* that are the subject of this right; or with respect to the *persons* allowed or forbidden to exercise it. And, in consequence of this authority, we find that the municipal laws of many nations have exerted such power of restraint; have in general forbidden the entering on another man's grounds, for any cause, without the owner's leave; have extended their protection to such particular animals as are usually the objects of pursuit; and have invested the prerogative of hunting and taking such animals in the sovereign of the state only, and such as he shall authorise.[2] Many reasons have concurred for making these constitutions: as, 1. For the encouragement of agriculture and improvement of lands, by giving every man an exclusive dominion over his own soil, 2. For preservation of the several species of these animals, which would soon be extirpated by a general liberty. 3. For prevention of idleness and dissipation in husbandsmen, artificers, and others of lower rank; which would he the unavoidable consequence of universal licence. 4. For prevention of popular insurrections and resistance to the government, by disarming the bulk of the people;[3] which last is a reason oftener meant, than avowed, by the makers of forest or game laws. Nor, certainly, in these prohibitions is there any *natural* injustice, as some have weakly enough supposed: since, as Puffendorf observes, the law does not hereby take from any man his present property, or what was already his own, but barely abridges him of one means of acquiring a future property, that of occupancy; which indeed the law of nature would allow him, but of which the laws of society have in most instances very justly and reasonably deprived him.

Yet, however defensible these provisions in general may be, on the footing of reason, or justice, or civil policy, we must notwithstanding acknowledge that, in their present shape, they owe their immediate original to slavery. It is not till after the irruption of the northern nations into the Roman empire, that we read of any other prohibitions, than that natural one of not sporting on any private grounds without the owner's leave; and another of a more spiritual nature, which was rather a rule of ecclesiastical discipline, than a branch of municipal law. The Roman or civil law, though it knew no restriction as to *persons* or *animals*, so far regarded the article of *place*, that it allowed no man to hunt or sport upon another's ground, but by consent of the owner of the soil. "*Qui alienum fundum ingreditur, venandi aut aucupandi gratia, potest a domino prohiberi ne ingrediatur.*"[4] For if there can, by the law of nature, be any inchoate imperfect property supposed in wild animals before they are taken, it seems most reasonable to fix it in him upon whose land they are found. And as to the other restriction, which relates to *persons* and not to *place*, the pontifical or canon law[5] interdicts "*venationes, et sylvaticas vagationes cum canibus et accipitribus*," to all *clergymen* without distinction; grounded on a saying of St. Jerom,[6] that it never is recorded that these diversions were used by the saints, or primitive fathers. And the canons of our Saxon church, published in the reign of king Edgar,[7] concur in the same prohibition: though our secular laws, at least after the conquest, did even in the times of popery dispense with this canonical impediment; and spiritual persons were allowed by the common law to hunt for their recreation, in order to render them fitter for the performance of their duty: as a confirmation whereof we may observe, that it is to this day a branch of the king's prerogative, at the death of every bishop, to have his kennel of hounds, or a composition in lieu thereof.[8]

But, with regard to the rise and original of our present civil prohibitions, it will be found that all forest and game laws were introduced into Europe at the same time, and by the same policy, as gave birth to the feodal system; when those swarms of barbarians issued from their northern hive, and laid the foundation of most of the present kingdoms of Europe, on the ruins of the western empire. For when a conquering general came to settle the oeconomy of a vanquished country, and to part it out among his soldiers or feudatories, who were to render him military service for such donations; it behoved him, in order to secure his new acquisitions, to keep the *rustici* or natives of the country, and all who were not his military tenants, in as low a condition as possible, and especially to prohibit them the use of arms. Nothing could do this more effectually than a prohibition of hunting and sporting; and therefore it was the policy of the conqueror to reserve this right to himself, and such on whom he should bestow it; which were only his capital feudatories, or greater barons. And accordingly we find, in the feudal constitutions,[9] one and the same law prohibiting the *rustici* in general from carrying arms, and also proscribing the use of nets, snares, or other engines for destroying the game. This exclusive privilege well suited the martial genius of the conquering troops, who delighted in a sport[10] which in it's pursuit and slaughter bore some resemblance to war. *Vita omnis,* (says Caesar, speaking of the antient Germans,) *in venationibus atque in studiis rei militaris consistit.*[11] And Tacitus in like manner observes, that *quoties bella non ineunt multum venatibus, plus per otium transigunt.*[12] And indeed, like some of their modern successors, they had no other amusement to entertain their vacant hours; despising all arts as effeminate, and having no other learning, than was couched in such rude ditties, as were sung at the solemn carousals which succeeded these antient huntings. And it is remarkable, that, in those nations where the feodal policy remains the most uncorrupted, the forest or game laws continue in their highest rigor. In France all game is properly the king's; and in some parts of Germany, it is death for a peasant to be found hunting in the woods of the nobility.[13,14]

With us in England also, hunting has ever been esteemed a most princely diversion and exercise. The whole island was replenished with all sorts of game in the times of the Britons; who lived in a wild and pastoral manner, without inclosing or improving their grounds, and derived much of their subsistence from the chace, which they all enjoyed in common. But when husbandry took place under the Saxon government, and lands began to be cultivated, improved, and enclosed, the beasts naturally fled into the woody and desart tracts; which were called the forests, and, having never been disposed of in the first distribution of lands, were therefore held to belong to the crown. These were filled with great plenty of game, which our royal sportsmen reserved for their own diversion, on pain of a pecuniary forfeiture for such as interfered with their sovereign. But every freeholder had the full liberty of sporting upon his own territories, provided he abstained from the king's forests: as is fully expressed in the laws of Canute,[15] and of Edward the Confessor:[16] "*sit quilibet homo dignus venatione sua, in sylva, et in agris, sibi propriis, et in dominio suo: et absineat omnis homo a venariis, regiis, ubicunque pacem eis habere voluerit*": which indeed was the antient law of the Scandinavian continent, from whence Canute probably derived it. "*Cuique enim in proprio fundo quamlibet feram quoquo modo venari permissum.*"[17]

However, upon the Norman conquest, a new doctrine took place; and the right of pursuing and taking all beasts of chase or *venary*, and such other animals as were accounted *game*, was then held to belong to the king, or to such only as were authorized under him. And this, as well upon the principles of the feodal law, that the king is the ultimate proprietor of all the lands in the kingdom, they being all held of him as the chief lord, or lord paramount of the fee; and, that therefore, he has the right of the universal soil, to enter thereon, and to chase and take such creatures at his pleasure: as also upon another maxim of the common law, which we have frequently cited and illustrated, that these animals are *bona vacantia*, and, having no other owner, belong to the king by his prerogative. As, therefore, the former reason was held to vest in the king a *right* to pursue and take them any where; the latter was supposed to give the king, and such as he should authorize, a *sole* and *exclusive* right.

This right, thus newly vested in the crown, was exerted with the utmost rigor, at and after the time of the Norman establishment; not only in the antient forests, but in the new ones which the conqueror made, by laying together vast tracts of country, depopulated for that purpose, and reserved solely for the king's royal diversion; in which were exercised the most horrid tyrannies and oppressions, under colour of forest law, for the sake of preserving the beasts of chase; to kill any of which, within the limits of the forest, was as penal as the death of a man. And, in pursuance of the same principle, king John laid a total interdict upon the *winged* as well as the *fourfooted* creation: "*capturam avium per totam Angliam interdixit.*"[18] The cruel and insupportable hardships, which these forest laws created to the subject, occasioned our ancestors to be as zealous for their reformation, as for the relaxation of the feodal rigors and the other exactions introduced by the Norman family; and, accordingly, we find the immunities of *carta de foresta* as warmly contended for, and extorted from the king with as much difficulty, as those of *magna carta* itself. By this charter, confirmed in parliament,[19] many forests were disafforested, or stripped of their oppressive privileges, and regulations were made in the regimen of such as remained: particularly[20] killing the king's deer was made no longer a capital offence, but only punished by a fine, imprisonment, or abjuration of the realm. And by a variety of subsequent statutes, together with the long acquiescence of the crown without exerting the forest laws, this prerogative is now become no longer a grievance to the subject.

But, as the king reserved to himself the *forests* for his own exclusive diversion, so he granted out from time to time other tracts of land to his subjects under the names of *chases* or *parks*,[21] or gave them licence to make such in their own grounds; which indeed are smaller forests, in the hands of a subject, but not governed by the forest laws: and by the common law no person is at liberty to take or kill any beasts of chase, but such as hath an antient chase or park; unless they be also beasts of prey.

As to all inferior species of game, called beasts and fowls of warren, the liberty of taking or killing them is another franchise or royalty, derived likewise from the crown, and called *free warren;* a word, which signifies preservation or custody: as the exclusive liberty of taking and killing fish in a public stream or river is called a *free fishery;* of which however no new franchise can at present be granted, by the express provision of *magna carta*, c. 16.[22] The principal intention of granting to any one these franchises or liberties was in order to protect the game, by giving the grantee a sole and exclusive power of killing it himself, provided he prevented other persons. And no man, but he who has a chase or free warren, by grant from the crown, or prescription which supposes one, can justify hunting or sporting upon another man's soil; nor indeed, in thorough strictness of common law, either hunting or sporting at all.

However novel this doctrine may seem, to such as call themselves *qualified* sportsmen, it is a regular consequence from what has been before delivered;

that the sole right of taking and destroying game belongs exclusively to the king. This appears, as well from the historical deduction here made, as because he may grant to his subjects an exclusive right of taking them; which he could not do, unless such a right was first inherent in himself. And hence it will follow, that no person whatever, but he who has such derivative right from the crown, is by common law entitled to take or kill any beasts of chase, or other game whatsoever. It is true, that, by the acquiescence of the crown, the frequent grants of free warren in antient times, and the introduction of new penalties of late by certain statutes for preserving the game, this exclusive prerogative of the king is little known or considered; every man, that is exempted from these modern penalties, looking upon himself as at liberty to do what he pleases with the game: whereas the contrary is strictly true, that no man, however well *qualified* he may vulgarly be esteemed, has a right to encroach on the royal prerogative by the killing of game, unless he can shew a particular grant of free warren; or a prescription, which presumes a grant: or some authority under an act of parliament. As for the latter, I recollect but two instances wherein an express permission to kill game was ever given by statute: the one by 1 Jac. I. cap. 27, altered by 7 Jac. I. cap. 11, and virtually repealed by 22 and 23 Car. II. c. 25, which gave authority, so long as they remained in force, to the owners of free warren, to lords of manors, and to all freeholders having 40*l. per annum* in lands of inheritance, or 80*l.* for life or lives, or 400*l.* personal estate (and their servants) to take partridges and pheasants upon their own, or their master's, free warren, inheritance, or freehold: the other by 5 Ann. c. 14, which empowers lords and ladies of manors to appoint gamekeepers to kill game for the use of such lord or lady; which with some alteration still subsists, and plainly supposes such power not to have been in them before. The truth of the matter is, that these game laws (of which we shall have occasion to speak again in the fourth book of these commentaries) do indeed *qualify* nobody, except in the instance of a gamekeeper, to kill game: but only, to save the trouble and formal process of an action by the person injured, who perhaps too might remit the offence, these statutes inflict *additional* penalties, to be recovered either in a regular or summary way, by any of the king's subjects, from certain persons of inferior rank who may be found offending in this particular. But it does not follow that persons, excused from these additional penalties, are, therefore, *authorised* to kill game. The circumstance of having 100*l. per annum*, and the rest, are not properly qualifications, but exemptions. And these persons, so exempted from the penalties of the game statutes, are not only liable to actions of trespass by the owners of the land; but also, if they kill game within the limits of any royal franchise, they are liable to the actions of such who may have the right of chase or free warren therein.

Upon the whole it appears, that the king, by his prerogative, and such persons as have, under his authority, the royal franchises of chase, park, free-warren, or free fishery, are the only persons who may acquire any property, however fugitive and transitory, in these animals *ferae naturae*, while living; which is said to be vested in them, as was observed in a former chapter, *propter privilegium.* And it must also be remembered, that such persons as may thus lawfully hunt, fish, or fowl, *ratione prvilegii,* have (as has been said) only a qualified property in these animals: it not being absolute or permanent, but lasting only so long as the creatures remain within the limits of such respective franchise or liberty, and ceasing the instant they voluntarily pass out of it. It is held, indeed, that if a man starts any game within his own grounds, and follows into another's, and kills it there, the property remains in himself.[23] And this is grounded on reason and natural justice:[24] for the property consists in the possession; which possession commences by the finding it in his own liberty, and is continued by the immediate pursuit. And so, if a stranger starts game in one man's chase or free-warren, and hunts it into another liberty, the property continues in the owner of the chase or warren; this property, arising from privilege,[25] and not being changed by the act of a mere stranger. Or if a man starts game on another's private grounds, and kills it there, the property belongs to him in whose ground it was killed, because it was also started there;[26] this property arising *ratione soli.* Whereas if, after being started there, it is killed in the grounds of a third person, the property belongs not to the owner of the first ground, because the property is local; nor yet to the owner of the second, because it was not started in his soil; but it vests in the person who started and killed it,[27] though guilty of a trespass against both the owners.

NOTES

1. Inst. 2, 1, 12.
2. Puff. L. N. 1.4. c. 6. Sec. 5.
3. Warburton's Alliance, 324.
4. Inst. 2. 1. Sec. 12.
5. Decretal. 1. 5. tit. 24. c. 2.
6. Decret. part.1 .dist: 34.1. 1.
7. Cap. 64.
8. 4 lnst. 3O9.
9. Feud. 1. 2. tit. 27. Sec. 5.
10. In the laws of Jenghiz Khan, founder of the Mogul and Tartarian empire, published A. D. 1205, there is one which prohibits the killing of all game from March to October; that the court and soldiery might find plenty enough in the winter, during their recess from war. (Mod. Univ. Hist. iv. 468.)
11. De Bell. Gall. l. 6. c. 20.
12. c. 15.
13. Mattheus de Crimin. c. 3. tit. 1. Carpzov. Practic. Saxonic. p. 2. c. 84.
14. An attentive perusal of the preceding pages must be sufficient to convince us, that the game-laws are among the powerful instruments of state-enginery, for the purpose of retaining the mass of the people in a state of the most abject subjection.

    The bill of rights, 1 W. and M, says Mr. Blackstone (Vol. 1. p. 143,) secures to the subjects of England the right of having arms for their defence, suitable to their condition and degree. In the construction of these game laws it seems to be held, that no person who is not qualified according to law to kill game, hath any right to keep a gun in his house. Now, as no person, (except the game-keeper of a lord or lady of a manor) is admitted to be qualified to kill game, unless he has 100l. per annum, &c. it follows that no others can keep a gun for their defence; so that the whole nation are completely disarmed, and left at the mercy of the government, under the pretext of preserving the breed of hares and partridges, for the exclusive use of the independent country gentlemen. In America we may reasonably hope that the people will never cease to regard the right of keeping and bearing arms as the surest pledge of their liberty [Tucker's note].

15. c. 77.
16. c. 36.
17. Stiernhoek de jure Sueon, l.2. c.8.
18. M. Paris, 303.
19. 9 Hen. III.
20. cap. 10.
21. See page 38.
22. Mirr. c. 5. Sec. 2. See p. 40.
23. 11 Mod. 75.
24. Puff. L. 1. 4. c. 6.
25. Lord Raym. 251.
26. Ibid.
27. Farr. 18. Lord Raym. Ibid.

2

William Penn (1644–1718)

Born to a wealthy Anglican family, William Penn was first attracted to Quakerism while a student at Oxford in the early 1660s. In the years following Oliver Cromwell's death in 1658 and the restoration of the English monarchy under Charles II, the Quakers were considered suspect by both Puritans and monarchists for various reasons including their pacifism and refusal to swear loyalty oaths. By the latter part of that decade, Penn had become closely linked with the sect and had written a number of treatises in support of Quaker doctrines and religious freedom generally. Beginning in 1667, Penn was arrested and tried several times for various offenses related to his espousal of Quakerism. The following decade, Penn worked with a group of other prominent Quakers to establish a colony in New Jersey, and was instrumental in drafting a charter of liberties for the settlement. In settlement of a substantial debt owed to his father, Charles II granted to Penn an enormous tract of land in the colonies, designating him as "Proprietary and Governor" of the new province of Pennsylvania in 1681. Penn spent 1682–1684 in Pennsylvania, and returned for another visit in 1699, but failing health and legal issues pertaining to his financial affairs forced him to remain in England for the remainder of his life.

Penn is well known for the vision of democracy and religious tolerance he exhibited in documents such as Concessions to the Province of Pennsylvania *(1681) and* Charter of Libertie *(1682), works that influenced the framers of the Constitution nearly one hundred years later. His* Concessions to the Province of Pennsylvania, *ratified by Penn and the members of the Provincial Council of Pennsylvania on July 11, 1681, is notable not only for the foresight of this planning document but also for the way in which fair dealing with native peoples and responsible stewardship of the land are included among the guiding principles of the colony.*

Concessions to the Province of Pennsylvania (1681)

CERTAIN CONDITIONS OR CONCESSIONS,

Agreed upon by William Penn, Proprietary and Governor of the Province of Pennsylvania and those who are the adventurers and purchasers in the same province the Eleventh of July, one thousand six hundred and eighty-one.

First. That so soon as it pleaseth God that the above said persons arrive there, a quantity of land or Ground plat shall be laid out for a large Town or City in the most convenient place upon the River for health and navigation; and every purchaser and adventurer shall by lot have so much land therein as will answer to the proportion which he hath bought or taken up upon rent. But it is to be noted that the surveyors shall consider what Roads or Highways will be necessary to the Cities, Towns, or through the lands. Great roads from City to City not to contain less than forty feet in breadth shall be first laid out and declared to be for highways before the Dividend of acres be laid out for the purchaser, and the like observation to be had for the streets in the Towns and Cities, that there may be convenient roads and streets preserved not to be encroached upon by any planter or builder that none may build irregularly to the damage of another. In this custom governs.

Secondly. That the land in the Town be laid out together after the proportion of ten thousand acres of the whole country, that is two hundred acres, if the place will bear it: However that the proportion be by lot and entire so as those that desire to be together, especially those that are by the catalogue laid together, may be so laid together both in the Town & Country.

Thirdly. That when the Country lots are laid out, every purchaser from one thousand to Ten thousand acres or more, not to have above One thousand acres together, unless in three years they plant a family upon every thousand acres; but that all such as purchase together, lie together; and if as many as comply with this Condition, that the whole be laid out together.

Fourthly. That where any number of purchasers, more or less, whose number of acres amounts to Five or ten thousand acres, desire to sit together in a lot or Township, they shall have their lot or Township cast together, in such places as have convenient Harbours or navigable rivers attending it, if such can be found, and in case any one or more Purchasers plant not according to agreement, in this concession to the prejudice of others of the same Township upon complaint thereof, made to the Governor or his deputy, with assistance they may award (if they see cause) that the complaining purchaser may, paying the survey money, and purchase money, and Interest thereof, be entitled, inrolled, and lawfully invested in the lands so not seated.

Fifthly. That the proportion of lands that shall be laid out in the first great Town or City, for every purchaser, shall be, after the proportion of Ten acres, for every Five hundred acres purchased, if the place will allow it.

Sixthly. That notwithstanding there be no mention made in the several Deeds made to the purchasers, yet the said William Penn, does accord and declare, that all Rivers, Rivulets, Woods and Underwoods, Waters, Watercourses, Quarries, Mines and Minerals, (except mines Royal) shall be freely and fully enjoyed and wholly by the purchasers into whose lot they fall.

Seventhly. That for every Fifty acres that shall be allotted to a servant, at the end of his service, his Quitrent shall be two shillings per annum, and the master or owner of the Servant, when he shall take up the other Fifty acres, his Quitrent shall be Four shillings by the year, or if the master of the servant, (by reason of the Indentures be is so obliged to do,) allot out to the Servant Fifty acres in his own Division, the said master shall have on demand allotted him from the Governor, the One hundred acres, at the chief rent of Six shillings per annum.

Eighthly. And for the encouragement of such as are ingenious, and willing to search out Gold and silver mines in this province, it is hereby agreed that they have liberty to bore and dig in any man's property, fully paying the damage done, and in case a Discovery should be made, that the discoverer have One Fifth, the owner of the soil (if not the Discoverer) a Tenth part, the Governor Two Fifths, and the rest to the public Treasury, saving to the king the share reserved by patent.

Ninthly. In every hundred thousand acres, the Governor and Proprietary by lot reserveth Ten to himself, which shall lie but in one place.

Tenthly. That every man shall be bound to plant or man so much of his share of Land as shall be set out and surveyed within three years after it is so set out and surveyed, or else it shall be lawful for new comers to be settled thereupon, paying to them their survey money, and they go up higher for their shares.

Eleventhly. There shall be no buying and selling, be it with an Indian, or one among another of any Goods to be exported but what shall be performed in public market, when such place shall be set apart or erected, where they shall pass the public Stamp or Mark. If bad ware and prized as good, or deceitful in proportion or weight, to forfeit the value as if good, and full weight and proportion to the public Treasury of the Province, whether it be the merchandize of the Indian or that of the Planters.

Twelfthly. And forasmuch as it is usual with the planters to overreach the poor natives of the Country in Trade, by Goods not being good of the kind, or debased with mixtures, with which they are sensibly aggrieved, it is agreed, whatever is sold to the Indians, in consideration of their furs, shall be sold in the market place, and there suffer the test, whether good or bad; if good to pass; if not good, not to be sold for good, that the natives may not be abused nor provoked.

Thirteenthly. That no man shall by any ways or means, in word or deed, affront or wrong any Indian, but he shall incur the same penalty of the Law, as if he had committed it against his fellow planters; and if any Indian shall abuse, in Word or Deed, any planter of this province, that he shall not be his own Judge upon the Indian, but he shall make his complaint to the Governor of the province, or his Lieutenant or Deputy, or some inferior magistrate near him, who shall, to the utmost of his power, take care with the king of the said Indian, that all reasonable Satisfaction be made to the said injured planter.

Fourteenthly. That all differences between the Planters and the natives shall also be ended by Twelve men, that is, by Six planters and Six natives, that so we may live friendly together as much as in us lieth, preventing all occasions of Heart burnings and mischief.

Fifteenthly. That the Indians shall have liberty to do all things relating to improvement of their Ground, and providing sustenance for the families, that any of the planters shall enjoy.

Sixteenthly. That the laws as to Slanders, Drunkenness, Swearing, Cursing, Pride in apparel, Trespasses, Distresses, Replevins, Weights and measures, shall be the same as in England, till altered by law in this province.

Seventeenthly. That all shall mark their hogs, sheep and other cattle, and what are not marked within three months after it is in their possession, be it young or old, it shall be forfeited to the Governor, that so people may be compelled to avoid the occasions of much strife between Planters.

Eighteenthly. That in clearing the ground, care be taken to leave One acre of trees for every five acres cleared, especially to preserve oak and mulberries, for silk and shipping.

Nineteenthly. That all ship masters shall give an account of their Countries, Names, Ships, Owners, Freights and Passengers, to an officer to be appointed for that purpose, which shall be registered within Two days after their arrival; and if they shall refuse so to do that then noe presume to trade with them, upon forfeiture thereof; and that such masters be looked upon as having an evil intention to the province.

Twentiethly. That no person leave the province without publication being made thereof in the market place, Three weeks before, and a certificate from some Justice of the peace, of his clearness with his neighbours and those he has dealt withal so far as such an assurance can be attained and given; and if any master of a ship shall contrary hereunto receive, and carry away any person that hath not given that public notice, the said master shall be liable to all debts owing by the said person, so secretly transported from the province. Lastly, that these are to be added to or corrected by and with the consent of the parties hereunto subscribed.

3

Benjamin Franklin (1706–1790)

Ben Franklin was apprenticed to his half-brother James, a Boston printer and editor of the New England Courant *at the age of twelve, and worked with him until he ran off to Philadelphia in 1723. Within a few years he had founded his own successful printing business in Philadelphia and had established a national reputation as the author/publisher of* Poor Richard's Almanack *(1733–1758) and numerous treatises on subjects including politics, science, and education. Franklin became one of Philadelphia's leading citizens, and was instrumental in the founding of the nation's first public library and numerous other philanthropic, educational, and civic projects. Franklin's interest in the natural sciences was lifelong and passionate. His work on electricity was published in Europe in 1751 to great acclaim, and he wrote prolifically on other scientific topics as well. He maintained a correspondence and friendship with numerous scientists and naturalists including fellow Philadelphians John Bartram (1699–1777) and his son William (1739–1823). Franklin's remarkable abilities and energy were such that even before the American Revolution and his crucial role as one of the drafters of the Declaration of Independence and wartime ambassador to France he had already secured a reputation as one of the leading figures of his age. Franklin began writing his* Autobiography *in 1771 and, although it was still far from complete at the time of his death, the work has gained a lasting literary reputation.*

Seventeenth-century colonists arriving from England, which had long since been largely deforested, were impressed by the ready availability of enormous quantities of firewood. As Francis Higginson wrote in New England's Plantation *(1630), the colonies made "good living for those that love good Fires." In heavily settled areas, however, such good living soon ended. By 1638, John Winthrop noted in his journal that the scarcity of firewood in Boston that winter had necessitated a wood-gathering trip to Spectacle Island in Massachusetts Bay; several members of the party suffered frostbite and one died from exposure. By the first half of the eighteenth century, the scarcity of firewood near large eastern cities, including Franklin's own Philadelphia, had resulted in price speculation and sporadic winter fuel shortages. In response to these developments, Franklin invented an energy-efficient stove that he called the "Pennsylvanian*

Fire-place." The civic-minded Franklin declined to take a patent out on his invention, telling Governor George Thomas, "That as we enjoy great advantages from the Invention of others, we should be glad of an Opportunity to serve others by any Invention of ours, and this we should do freely and generously."

From *An Account of the Newly Invented Pennsylvanian Fire-places* (1744)

In these Northern Colonies the Inhabitants keep FIRES to sit by, generally *Seven Months* in the Year; that is, from the Beginning of *October* to the End of *April;* and in some Winters near *Eight Months*, by taking in part of September and May.

WOOD, our common Fewel, which within these 100 Years might be had at every Man's Door, must now be fetch'd near 100 Miles to some Towns, and makes a very considerable Article in the Expence of Families.

As therefore so much of the Comfort and Conveniency of our Lives, for so great a Part of the Year, depends on the Article of FIRE;—since Fuel is become so expensive, and (as the Country is more clear'd and settled) will of course grow scarcer and dearer; any new Proposal for Saving the Wood, and for lessening the Charge and augmenting the Benefit of FIRE, by some particular Method of Making and Managing it, may at least be thought worth Consideration.

THE NEW FIRE-PLACES are a late Invention to that purpose (experienced now three Winters by a great Number of Families in *Pennsylvania)* of which this Paper is intended to give a particular Account.

* * *

THE ADVANTAGES OF THIS FIRE-PLACE

Its Advantages above the common FirePlaces are,

1. That your whole Room is equally warmed; so that People need not croud so close round the Fire, but may sit near the Window and have the Benefit of the Light for Reading, Writing, Needle-work, *&c.* They may sit with Comfort in any Part of the Room; which is a very considerable Advantage in a large Family, where there must often be two Fires kept, because all cannot conveniently come at One.
2. If you sit near the Fire, you have not that cold Draught of uncomfortable Air nipping your Back and Heels, as when before common Fires, by which many catch Cold, being scorcht before and as it were froze behind.
3. If you sit against a Crevice, there is not that sharp Draught of cold Air playing on you, as in Rooms where there are Fires in the common way; by

which many catch Cold, whence proceed Coughs,[1] Catarrhs, Tooth-Achs, Fevers, Pleurisies and many other Diseases.

4. In Case of Sickness, they make most excellent Nursing-Rooms; as they constantly supply a Sufficiency of fresh Air, so warmed at the same time as to be no way inconvenient or dangerous. A small One does well in a Chamber; and, the Chimneys being fitted for it, it may be remov'd from one Room to another as Occasion requires, and fix'd in half an Hour. The equal Temper, too, and Warmth, of the Air of the Room, is thought to be particularly advantageous in some Distempers: For 'twas Observ'd in the Winters of 1730 and 1736, when the Small-Pox spread in *Pennsylvania*, that very few of the Children of the *Germans* died of that Distemper, in Proportion to those of the *English*; which was ascrib'd by some to the Warmth and equal Temper of Air in their Stove-Rooms; which made the Disease as favourable as it commonly is in the *West Indies*. But this Conjecture we submit to the Judgment of Physicians.
5. In common Chimneys the strongest Heat from the Fire, which is upwards, goes directly up the Chimney, and is lost; and there is such a strong Draught into the Chimney, that not only the upright Heat, but also the back, sides and downward Heats, are carried up the Chimney by that Draught of Air; and the Warmth given before the Fire by the Rays that strike out towards the Room, is continually driven back, crouded into the Chimney, and carried up, by the same Draught of Air.—But here the upright Heat, strikes and heats the Top Plate, which warms the Air above it, and that comes into the Room. The Heat likewise which the Fire communicates to the Sides, Back Bottom and Air-Box, is all brought into the Room; for you will find a constant Current of warm Air coming out of the Chimney-Corner into the Room. Hold a Candle just under the Mantle-Piece or Breast of your Chimney, and you will see the Flame bent outwards: By laying a Piece of smoaking Paper on the Hearth, on either Side, you may see how the Current of Air moves, and where it tends, for it will turn and carry the Smoke with it.
6. Thus as very little of the Heat is lost, when this Fire-Place is us'd, *much less Wood*[2] will serve you, which is a considerable Advantage where Wood is dear.
7. When you burn Candles near this Fire-Place, you will find that the Flame burns quite upright, and does not blare and run the Tallow down, by drawing towards the Chimney, as against common Fires.
8. This Fire-place cures most smoaky Chimneys, and thereby preserves both the Eyes and Furniture.
9. It prevents the Fouling of Chimneys; much of the Lint and Dust that contributes to foul a Chimney, being by the low Arch oblig'd to pass thro' the Flame, where 'tis consum'd. Then, less Wood being burnt, there is less Smoke made. Again, the Shutter, or Trap-Bellows, soon blowing the Wood into a Flame, the same Wood does not yield so much Smoke as if burnt in a common Chimney: For as soon as Flame begins, Smoke, in proportion, ceases.

10. And if a Chimney should be foul, 'tis much less likely to take Fire. If it should take Fire, 'tis easily stifled and extinguished.
11. A Fire may be very speedily made in this Fire-Place, by the Help of the Shutter, or Trap-Bellows, as aforesaid.
12. A Fire may be soon extinguished, by closing it with the Shutter before, and turning the Register behind, which will stifle it, and the Brands will remain ready to rekindle.
13. The Room being once warm, the Warmth may be retain'd in it all Night.
14. And lastly, the Fire is so secur'd at Night, that not one Spark can fly out into the Room to do Damage.

With all these Conveniencies, you do not lose the pleasant Sight nor Use of the Fire, as in the Dutch Stoves, but may boil the Tea-Kettle, warm the Flat-Irons, heat Heaters, keep warm a Dish of Victuals by setting it on the Top, *&c. &c. &c.*

OBJECTIONS ANSWERED.

There are some Objections commonly made by People that are unacquainted with these Fire-Places, which it may not be amiss to endeavour to remove, as they arise from Prejudices which might otherwise obstruct in some Degree the general Use of this beneficial Machine. We frequently hear it said, *They are of the Nature of the Dutch Stoves; Stoves have an unpleasant Smell; Stoves are unwholesome;* and, *Warm Rooms make People tender and apt to catch Cold.*—As to the first, that they are of the Nature of *Dutch* Stoves, the Description of those Stoves in the Beginning of this Paper, compar'd with that of these Machines, shows that there is a most material Difference, and that these have vastly the Advantage, if it were only in the single Article of the Admission and Circulation of fresh Air. But it must be allowed there has been some Cause to complain of the offensive Smell of Iron Stoves. This Smell, however, never proceeded from the Iron itself, which in its Nature, whether hot or cold, is one of the sweetest of Metals, but from the general uncleanly Manner of using those Stoves. If they are kept clean, they are as sweet as an Ironing-Box, which, tho' ever so hot, never offends the Smell of the nicest Lady: But it is common, to let them be greased by setting Candlesticks on them, or otherwise; to rub greasy Hands on them, and, above all, to spit upon them to try how hot they are, which is an inconsiderate, filthy unmannerly Custom; for the slimy Matter of Spittle drying on, burns and fumes when the Stove is hot, as well as the Grease, and smells most nauseously; which makes such close Stove-Rooms, where there is no Draught to carry off those filthy Vapours, almost intolerable to those that are not from their Infancy accustomed to them. At the same time, nothing is more easy than to keep them clean; for when by any Accident they happen to be fouled, a Lee made of Ashes, and Water, with a Brush, will scour them perfectly; as will also a little strong Soft-Soap and Water.

That hot Iron of itself gives no offensive Smell, those know very well, who have (as the Writer of this has) been present at a Furnace, when the Workmen were pouring out the flowing Metal to cast large Plates, and not the least Smell of it to be perceived. That hot Iron does not, like Lead, Brass, and some other Metals give out unwholesome Vapours, is plain from the general Health and Strength of those who constantly work in Iron as Furnace-men, Forge-Men, and Smiths; That it is in its Nature a Metal perfectly wholesome to the Body of Man, is known from the beneficial Use of Chalybeat or Iron-Mine Waters; from the Good done by taking Steel Filings in several Disorders; and that even the Smithy Water in which hot Irons are quench'd, is found advantageous to the human Constitution.—The ingenious and learned Dr. *Desaguliers*, to whose instructive Writings the Contriver of this Machine acknowledges himself much indebted, relates an Experiment he made, to try whether heated Iron would yield unwholesome Vapours; He took a Cube of Iron, and having given it a very great Heat, he fix'd it to a Receiver, exhausted by the Air Pump, that all the Air rushing in to fill the Receiver, should first pass thro' a Hole in the hot Iron. He then put a small Bird into the Receiver, who breath'd that Air without any Inconvenience or suffering the least Disorder. But the same Experiment being made with a Cube of hot Brass, a Bird put into that Air dy'd in a few Minutes. Brass indeed stinks even when cold, and much more when hot; Lead too, when hot, yields a very unwholesome Steam; but IRON is always sweet, and every way taken is wholesome and friendly to the human Body—except in Weapons.

That warm Rooms make People tender and apt to catch Cold, is a Mistake as great as it is (among the *English*) general. We have seen in the preceding Pages how the common Rooms are apt to give Colds; but the Writer of this Paper may affirm, from his own Experience, and that of his Family and Friends who have used warm Rooms for these four Winters past, that by the Use of such Rooms, People are rendered *less liable* to take Cold, and indeed *actually hardened*. If sitting warm in a Room made One subject to take Cold on going out, lying warm in Bed should, by a Parity of Reason, produce the same Effect when we rise; Yet we find we can leap out of the warmest Bed naked in the coldest Morning, without any such Danger; and in the same Manner out of warm Clothes into a cold Bed. The Reason is, that in these Cases the Pores all close at once, the Cold is shut out, and the Heat within augmented, as we soon after feel by the glowing of the Flesh and Skin. Thus no one was ever known to catch Cold by the Use of the Cold Bath: And are not cold Baths allowed to harden the Bodies of those that use them? Are they not therefore frequently prescrib'd to the tenderest Constitutions? Now every Time you go out of a warm Room into the cold freezing Air, you do as it were plunge into a Cold Bath, and the Effect is in proportion the same; for (tho' perhaps you may feel somewhat chilly at first) you find in a little Time your Bodies hardened and strengthened, your Blood is driven round with a brisker Circulation, and a comfortable steady uniform inward Warmth succeeds that equal outward Warmth you first received in the Room.

Farther to confirm this Assertion, we instance the *Swedes*, the *Danes*, the *Russians:* These Nations are said to live in Rooms, compar'd to ours, as hot as Ovens;[3] yet where are the hardy Soldiers, tho' bred in their boasted cool Houses, that can, like these People, bear the Fatigues of a Winter Campaign in so severe a Climate, march whole Days to the Neck in Snow, and at Night entrench in Ice, as they do?

The Mentioning of those Northern Nations puts me in Mind of a considerable *Publick Advantage* that may arise from the general *Use* of these Fire-places. It is observable, that tho' those Countries have been well inhabited for many Ages; Wood is still their Fuel, and yet at no very great Price; which could not have been if they had not universally used Stoves, but consum'd it as we do, in great Quantities by open Fires. By the Help of this saving Invention, our Wood may grow as fast as we consume it, and our Posterity may warm themselves at a moderate Rate, without being oblig'd to fetch their Fuel over the *Atlantick*; as, if Pit-Coal should not be here discovered (which is an Uncertainty) they must necessarily do.

We leave it to the *Political Arithmetician* to compute, how much Money will be sav'd to a Country; by its spending two thirds less of Fuel; how much Labour sav'd in Cutting and Carriage of it; how much more Land may be clear'd for Cultivation; how great the Profit by the additional Quantity of Work done, in those Trades particularly that do not exercise the Body so much, but that the Workfolks are oblig'd to run frequently to the Fire to warm themselves: And to Physicians to say, how much healthier thick-built Towns and Cities will be, now half suffocated with sulphury Smoke, when so much less of that Smoke shall be made, and the Air breath'd by the Inhabitants be consequently so much purer.

* * *

Notes

1. My Lord *Molesworth*, in his Account of *Denmark*, says, That "few or none of the People there, are troubled with Coughs, Catarrhs, Consumptions, or such like Diseases of the Lungs; so that in the Midst of Winter in the Churches, which are very much frequented, there is no Noise to interrupt the Attention due to the Preacher. I am persuaded (says he) their *warm Stoves* contribute to their Freedom from these kind of Maladies." Page 91.
2. People who have us'd these Fire-places, differ much in their Accounts of the Wood saved by them. Some say five sixths, others three fourths, and others much less. This is owing to the great Difference there was in their former Fires; some (according to the different Circumstances of their Rooms and Chimneys) having been us'd to make very large, others middling, and others, of a more sparing Temper, very small Ones: While in these Fire-places, (their Size and Draught being nearly the same) the Consumption is more equal. I suppose, taking a Number of Families together, that two thirds, or half the Wood at least, is saved. My common Room, I know, is made twice as warm as it used to be, with a quarter of the Wood I formerly consum'd there.

3. Mr. *Boyle*, in his Experiments and Observations upon cold, *Shaw's Abridgment, Vol.* I. page 684, says, " 'Tis remarkable, that while the Cold has strange and tragical Effects at *Moscow*, and elsewhere, the *Russians* and *Livonians* should be exempt from them, who accustom themselves to pass immediately from great Degree of Heat, to as great an one of cold, without receiving any visible Prejudice thereby. I remember being told by a person of unquestionable Credit, that it was a common Practice among them, to go from a hot Stove into cold Water; the same was, also, affirmed to me, by another who resided at *Moscow*. This Tradition is likewise abundantly confirmed by *Olearius*. " *'Tis a surprizing thing*," says he, "*to see how far the Russians can endure Heat; and how, when it makes them ready to faint, they can go out of their Stoves, stark naked, both Men and Women, and throw themselves into cold Water; and even in Winter wallow in the Snow.*"

4

Peter Kalm (1716–1779)

While a graduate student at Uppsala University in Sweden, Pehr (Peter) Kalm became a pupil of famed botanist and taxonomist Carl von Linné (Carolus Linnæus). In 1744 Kalm assisted Linné on a research trip in Russia, and the following year he was elected to the Swedish Academy of Sciences. In 1747 he was offered a professorship in agriculture at the Åbo Academy in what is today Finland, but was almost immediately given leave to travel to North America on a research trip sponsored by the Academy of Sciences. After a remarkably difficult trip that included a shipwreck and numerous delays, Kalm finally reached Philadelphia on September 15, 1848. Over the next three years, he traveled throughout the mid-Atlantic region (primarily Pennsylvania, New York, and New Jersey) and Canada. He returned to Sweden late in the spring of 1751, where he resumed his academic career and worked on the account of his Travels, *which appeared in three volumes (1753–1761). In addition to his work as a naturalist, Kalm was ordained as a Lutheran clergyman in 1757 and awarded a doctorate in theology in 1768 from the University of Lund.*

The specific objective of the Swedish Academy's proposed expedition to North America was to discover useful plants and trees that might be suitable for growth in the rigorous climate of Sweden. Kalm admirably fulfilled this part of his charge, returning to Sweden with numerous plant specimens for cultivation and study, but his Travels *also reflects Kalm's perceptive observations of cultural mores (despite a certain degree of parochialism) as well as the geography and natural history of the North American colonies. While in Philadelphia, Kalm benefited greatly from the friendship and expertise of the American naturalist John Bartram (1699–1777), and noted with approval Benjamin Franklin's recent invention, the Pennsylvanian Fire-Place. Like Franklin, Kalm had a strong utilitarian tendency, and his critical comments on colonial agriculture reflect his disdain for the wasteful practices he saw throughout the colonies: "In a word, the grain fields, the meadows, the forests, the cattle, etc. are treated with equal carelessness . . . their eyes are fixed upon the present gain, and they are blind to the future."*

From *Travels Into North America* (1753)

DECEMBER THE 8TH

The Newly invented Pennsylvania "Fireplaces." Although this city [of Philadelphia] is situated at 40° north latitude and should be relatively warm in winter, when we compare it with European towns of the same latitude, experience has taught that sometimes it is as cold here as in old Sweden. It is necessary only to examine the meteorological observations in this travelogue for this year, the supplementary remarks here and there in the diary about weather conditions and temperature, and what Mr. Franklin records in his *Poor Richard's Almanac Improved.* It has therefore been imperative to make some provisions for heating the houses for a period of several months in winter, and for that purpose various types of stoves have been used. I shall not attempt to describe them all at this time, because they are not only described but their good features and serious faults elaborated in the book which Mr. Franklin published in Philadelphia, 1744, under the name of *An Account of the New Invented Pennsylvanian Fire-Places* etc. But since all these had their faults, and the Englishmen liked to see the fire burn instead of confining it in a stove, Benjamin Franklin invented a new type of stove, which not only provides plenty of heat, saves fuel and brings fresh air into the room, but is so constructed that the flame may be seen. An extensive description of it is found in the above-mentioned book by Mr. Franklin, so that I have but little to add, especially since Mr. [Lewis] Evans has in addition made copious marginal notes about it in his presentation copy of it to me [and which may be consulted]. Mr. Franklin invented the stove and Mr. Evans made the drawings and figures. The bottom should not rest flat on the floor or ground, but be elevated a little, and an opening or passage left to the air-box, so that the air which is heated beneath the bottom plate can get out into the room. There are several types of this stove: some have dampers, which in the description are called registers, and others have not; some have a front plate near the bottom which can be moved up and down, and when it is moved down the draft becomes stronger and the wood begins to burn quicker; others have a hole on the frontal plate with a small trap-door which when opened allows air to enter and fan the fire, for there is a narrow passage leading to this trap-door, either from the room below or from a space under the floor connected with the outside. Under the air-box is an opening through which fresh, cold air enters it, and here it is heated and sent into the room through side openings. If there is a cellar beneath the room with the stove, a hole is made to it, so that the air can come from that direction; but otherwise it must pass under the floor to the air-box from some side of the house, preferably an entrance hall, where a small opening may be made in the wall near the floor, through which fresh air may enter. Several who could not afford to purchase these stoves imitated them by making them either of brick or white Dutch tile [or brick], making only the top of iron. But these were not so warm.

DECEMBER THE 9TH

More about the Franklin Stoves. I wrote yesterday about the newly invented fireplaces [Stoves]. I shall now add a few more details. They are made or cast of iron, i.e. the iron which is obtained in this province. The size varies and so does the price.

Despite their usefulness they have been criticised. It was held that if the chimney could not be swept [as in big chimneys with open fireplaces] there was danger of fire; some thought the stoves gave too much heat, and since the Englishmen were not accustomed to this they liked open fires better; and the Germans preferred *their* small, oblong, square iron stoves, which are constructed in the same way as the stoves in Bohuslän and Norway, because they give more heat and cost less. Nevertheless, there were many who used these new stoves, both in Philadelphia, New York and elsewhere. In the country where they had plenty of fuel they used the large fireplaces, since many believed these new ones expensive, and did not reckon the cost of wood. Mr. Franklin loaned me one of the stoves for the winter. It kept the house quite warm, but then one had to use short wood in it. It proved often unnecessary to have a fire in the kitchen, and one could prepare chocolate and other food in the little stove. Also, it proved possible, by suspending a cord from above in front of the fire-box, to roast meat or fowl attached to it and turned. And curiously enough it was roasted better and quicker than in a regular kitchen fireplace, since the room was warm and heat came both from the fire and the iron of the stove. The chimney is seldom cleaned more than once a year, but Mr. Franklin was in the habit of setting fire to a sheet of paper every fortnight and let it pass through the flue leading to the stove and so burn of the soot there also. If the stove is narrow it is not so easy to sweep the chimney, after everything is closed up by masonry; but Mr. Franklin had a brick removed beside the stove, let a man pass down through the chimney, clean it, and when he reached the bottom near the stove had him force the soot through the hole made by the removal of the brick. When this was done the brick was replaced. Where the hearth is broad the stove is placed on one side of it, and a door made on the other through which the chimney sweep can enter and do his work. To get fresh air into the stove of a house with no cellar, and where no outside air is wanted, Mr. Franklin this year had had the stove in his own room set on a rim of masonry six inches from the floor, with an opening through the bricks on one side to let the cold air near the floor enter, pass through the air-box, where it was heated, and then pass through the holes on the iron sides of the stove into the room, etc. This brought about a constant circulation of air.

OCTOBER THE 11TH

The Apple Crop. I have already mentioned, that every countryman has a greater or lesser number of apple trees planted round his farmhouse from which he gets large quantities of fruit, a part of which he sells. From another part he makes cider, and some are used in his own family for pies, tarts and the like. However, he cannot expect an equal amount of fruit every year, and I was told that this season had not by far yielded such a quantity of apples as the preceding, the cause being the continual severe drought in the month of May which had hurt all the blossoms of the apple trees and made them wither. The heat had been so great that it dried up all the plants and the grass in the fields.

The *Polytrichum commune.*, a species of moss, grew plentifully on wet and low meadows between the woods, and in several places quite covered them, as our mosses cover the meadows in Sweden. It was likewise very plentiful on hills.

Agriculture was in a very bad state hereabouts. Formerly when a person had bought a piece of land, which perhaps had never been plowed since Creation, he cut down a part of the wood, tore up the roots, tilled the ground, sowed seed on it, and the first time he got an excellent crop.—But the same land after being cultivated for several years in succession, without being manured, finally loses its fertility of course. Its possessor then leaves it fallow and proceeds to another part of his land, which he treats in the same manner. Thus he goes on till he has changed a great part of his possessions into grain fields, and by that means deprived the ground of its fertility. He then returns to the first field, which now has pretty well recovered. This he tills again as long as it will afford him a good crop; but when its fertility is exhausted he leaves it fallow again and proceeds to the rest as before.

Careless Farming. It being customary here to let the cattle go about the fields and in the woods both day and night, the people cannot collect much dung for manure. But by leaving the land fallow for several years a great quantity of weeds spring up in it, and get such strength that it requires a considerable time to extirpate them. This is the reason why the grain is always so mixed with the seed of weeds. The great richness of the soil which the first European colonists found here, and which had never been plowed before, has given rise to this neglect of agriculture, which is still observed by many of the inhabitants. But they do not consider that when the earth is quite exhausted a great space of time and an infinite deal of labor are necessary to bring it again into good condition, especially in these countries which are almost every summer scorched by the excessive heat and drought. The soil of the grain fields consists of a thin mould, greatly mixed with a brick-colored sand and clay and a quantity of small particles of glimmer (mica). The latter comes from stones which are found here almost everywhere at the depth of a foot or so. These little pieces of mica make the ground sparkle when the sun shines upon it.

MAY THE 18TH

Sweden versus America. Though it was already quite late in May the nights were very dark here. About an hour after sunset it was so dark that it was impossible to read a book, though the type was ever so large. About ten o'clock on a clear night the darkness had increased so much that it looked like one of the darkest though starlight nights in an autumn in Sweden. It also seemed to me that though the nights were clear, the stars did not give so great a light as they did in Sweden. And as at this season the nights were usually dark, and the sky covered with clouds, I could compare them only to dark and cloudy Swedish winter nights. It was therefore, at this time of the year, very difficult to travel during such cloudy nights; for neither man nor horse could find his way. The nights here in general seem very disagreeable to me, in comparison with the light and glorious summer nights of Sweden. Ignorance sometimes makes us think slightly of our native land, Sweden. If other countries have their advantages, Sweden also has hers; and upon comparing the advantages and disadvantages of different places, Sweden will be found to be not inferior to any of them.

Old and New Sweden Compared. I shall briefly mention in what points I think Sweden is preferable to this part of America; and why I, as Ulysses did with Ithaca, prefer Old Sweden to New Sweden.

Dark Nights and Changeable Weather. The nights are very dark here all summer; and in winter they are quite as dark, if not darker, than the winter nights in Sweden; for here is no Aurora Borealis, and the stars give a very faint light. It is very remarkable if an Aurora Borealis appears once or twice a year. The winters here bring no snow, to make the nights clear and travelling more safe and easy. The cold is, however, frequently as intense as in Old Sweden. The snow which falls lies only a few days, and always goes off with a great deal of wet. The rattlesnakes, horned snakes, red-bellied, green, and other poisonous snakes, against whose bite there is frequently no remedy, are numerous here. To these I must add the wood lice with which the forests are so pestered that it is impossible to pass through a bush or to sit down, though the place be ever so pleasant, without having a whole swarm of them on your clothes. The inconvenience and trouble they cause, both to man and beast, I have described in the *Memoirs* of the Royal Swedish Academy of Sciences. The weather is so inconstant here that when a day is most excessively hot, the next is often cold. This sudden change often happens in one day; and few people can endure these changes, without impairing their health. The heat in summer is intense, and the cold in winter often more piercing. However, one can always secure one's self against the cold; but when the great heat is of any duration, there is hardly any protection against it. It tires one so that one does not know which way to turn. It has frequently happened that people who walked into the fields dropped down dead on account of the violence of the heat. Several illnesses prevail here; and they increase every year. Nobody is left unattacked by the intermitting fever; and many people are forced to suffer it every year, together with other diseases. Peas cannot be sown, on account of the insects which consume them. There are worms in the rye seed, and myriads of them in the cherry trees. The caterpillars often eat all the leaves from the trees, so that they cannot bear fruit that year; and large numbers die every year, both of fruit trees and forest trees. The grass in the meadows is likewise consumed by a kind of worm; another species causes the plums to drop before they are half ripe. The oak here affords not nearly so good timber as the European oak. The fences cannot last above eighteen years. The houses are of no long duration. The meadows are poor, and what grass they have is bad. The pasture for cattle in the forests consists of such plants as they do not like, and which they are compelled to eat by necessity; for it is difficult to find any grass in the great forests where the trees stand far apart, and where the soil is excellent. For this reason, the cattle are forced, during almost the whole winter and part of the summer, to live upon the young shoots and branches of trees, which sometimes have no leaves: therefore, the cows give very little milk, and decrease in size every generation. The houses are extremely unfit for winter habitation. Hurricanes are frequent, which overthrow trees, carry away roofs, and sometimes houses, and do a great deal of damage. Some of these inconveniences might be remedied by diligence; but others cannot, or only with difficulty. Thus every country has its advantages, and its defects: happy is he who can content himself with his own lot!

Careless Agriculture. The rye grows very poorly in most of the fields, which is chiefly owing to the carelessness in agriculture, and to the poorness of the fields, which are seldom or never manured. After the inhabitants have converted a tract of land into a tillable field, which has been a forest for many centuries, and which consequently has a very fine soil, the colonists use it as such as long as it will bear any crops; and when it ceases to bear any, they turn it into pastures for the cattle, and take new grain fields in another place, where a rich black soil can be found and where it has never been made use of. This kind of agriculture will do for a time; but it will afterwards have bad consequences, as every one may clearly see. A few of the inhabitants, however, treated their fields a little better: the English in general have carried agriculture to a higher degree of perfection than any other nation. But the depth and richness of the soil found here by the English settler (as they were preparing land for plowing, which had been covered with woods from times immemorial) misled them, and made them careless husbandmen. It is well known that the Indians lived in this country for several centuries before the Europeans came into it; but it is likewise known, that they lived chiefly by hunting and fishing, and had hardly any agriculture. They planted corn and some species of beans and pumpkins; and at the same time it is certain that a plantation of such vegetables as serve an Indian family during one year take up no more ground than a farmer in our country takes to plant cabbage for his family. At least, a farmer's cabbage and turnip ground, taken together, is always as extensive, if not more so, than all the corn fields and kitchen gardens of an Indian family. Therefore, the Indians could hardly subsist for one month upon the produce of their gardens and fields. Commonly, the little villages of Indians are about twelve or eighteen miles distant from each other. Hence one may judge how little ground was formerly employed for planting; and the rest was overgrown with large, tall trees. And though they cleared (as is yet usual) new ground, as soon as the old one had lost its fertility, such little pieces as they made use of were very inconsiderable, when compared to the vast forest which remained. Thus the upper fertile soil increased considerably, for centuries; and the Europeans coming to America found a rich, fine soil before them, lying as loose between the trees as the best bed in a garden. They had nothing to do but to cut down the wood, put it up in heaps, and to clear the dead leaves away. They could then immediately proceed to plowing, which in such loose ground is very easy; and having sown their grain, they got a most plentiful harvest. This easy method of getting a rich crop has spoiled the English and other European settlers, and induced them to adopt the same method of agriculture as the Indians; that is, to sow uncultivated grounds, as long as they will produce a crop without manuring, but to turn them into pastures as soon as they can bear no more, and to take on new spots of ground, covered since ancient times with woods, which have been spared by the fire or the hatchet ever since the Creation. This is likewise the reason why agriculture and its science is so imperfect here that one can travel several days and learn almost nothing about land, neither from the English, nor from the Swedes, Germans, Dutch and French; except that from their gross mistakes and carelessness of the future, one finds opportunities every day of making all sorts of observations, and of growing wise by their mistakes. In a word, the grain fields, the meadows, the forests, the cattle, etc. are treated with equal carelessness; and the characteristics of the English nation, so well skilled in these branches of husbandry, is scarcely recognizable here. We can hardly be more hostile toward our woods in Sweden and Finland than they are here: their eyes are fixed upon the present gain, and they are blind to the future. Their cattle grow poorer daily in quality and size because of hunger, as I have before mentioned. On my travels in this country I observed several plants, which the horses and cows preferred to all others. They were wild in this country and likewise grew well on the driest and poorest ground, where no other plants would succeed. But the inhabitants did not know how to turn this to their advantage, owing to the little account made of Natural History, that science being here (as in other parts of the world) looked upon as a mere trifle, and the pastime of fools. I am certain, and my certainty is founded upon experience, that by means of these plants, in the space of a few years, I should be able to turn the poorest ground, which would hardly afford food for a cow, into the richest and most fertile meadow, where great flocks of cattle would find superfluous food, and grow fat. I own that these useful plants are not to be found on the grounds of every planter: but with a small share of natural knowledge, a man could easily collect them in the places where they are to be had. I was astonished when I heard the country people complaining of the badness of the pastures; but I likewise perceived their negligence, and often saw excellent plants growing on their own grounds, which only required a little more attention and assistance from their unexperienced owner. I found everywhere the wisdom and goodness of the Creator; but too seldom saw any inclination to make use of them or adequate estimation of them, among men.

O fortunatos nimium, sua si bona norint,
Agricolas!

Virg.

I have been led to these reflections, which may perhaps seem foreign to my purpose, by the bad and neglected state of agriculture in every part of this continent, and because I wanted to show the reason why this journal is so thinly stocked with economical advantages in several branches of husbandry. [There were so few of them]. I do not however deny, that I have here and there found skilful farmers, but they were very scarce.

5

THOMAS JEFFERSON (1743–1826)

Although Thomas Jefferson made his reputation as a statesman, he was fond of saying that nature had destined him for the sciences. Even during his tumultuous career in politics, Jefferson maintained his involvement in philosophy and the sciences, serving simultaneously as president of the American Philosophical Society and president of the United States, and playing a leading role in the planning and sponsorship of the Lewis and Clark expedition, which he saw as one of the great national achievements of his presidency.

Jefferson's only book-length work, Notes on the State of Virginia *(1787) reflects the enormous breadth of Jefferson's interests, including his extensive knowledge of the natural history of the state. He counters Peter Kalm's comments on the wastefulness of American agricultural practices, arguing: "The indifferent state of [agriculture] among us does not proceed from a want of knowledge merely; it is from our having such quantities of land to waste as we please. In Europe the object is to make the most of their land, labour being abundant; here it is to make the most of our labour, land being abundant." He also takes pains to rebut the assertions of the French naturalist Buffon that animals common to both the old world and the new were "degenerated" in America, pointing out that the fossilized remains of mammoths recently discovered in North America were "the largest of all terrestrial beings." Jefferson's nationalistic pride in the natural history of the land anticipates American literary nationalism and the cult of the American wilderness fostered by American landscape painters and writers of the early nineteenth century.*

From *Notes on the State of Virginia* (1787)

The opinion advanced by the Count de Buffon,[1] is 1. That the animals common both to the old and new world, are smaller in the latter. 2. That those peculiar to the new are on a smaller scale. 3. That those which have been domesticated in both, have degenerated in America: and 4. That on the whole it exhibits fewer species. And the reason he thinks is, that the heats of

America are less; that more waters are spread over its surface by nature, and fewer of these drained off by the hand of man. In other words, that *heat* is friendly, and *moisture* adverse to production and development of large quadrupeds. I will not meet this hypothesis on its first doubtful ground, whether the climate of America be comparatively more humid? Because we are not furnished with observations sufficient to decide this question. And though, till it be decided, we are as free to deny, as others are to affirm the fact, yet for a moment let it be supposed. The hypothesis after this supposition, proceeds to another; that *moisture* is unfriendly to animal growth. The truth of this is inscrutable to us by reasonings a priori. Nature has hidden from us her modus agendi. Our only appeal on such questions is to experience; and I think that experience is against the supposition. It is by the assistance of *heat* and *moisture* that vegetables are elaborated from the elements of earth, air, water, and fire. We accordingly see the more humid climates produce the greater quantity of vegetables. Vegetables are mediately or immediately the food of every animal: and in proportion to the quantity of food, we see animals not only multiplied in their numbers, but improved in their bulk, as far as the laws of their nature will admit. Of this opinion is the Count de Buffon himself in another part of his work:[2] "*en general il parois ques les pays un peu froids conviennent mieux à nos boeufs que les pays chauds, et qu'ils sont d'autant plus gros et plus grande que le climat est plus* humide *et plus abondans en paturages. Les boeufs de Danemarck, de la Podolie, de l'Ukraine et de la Tartarie qu'habitent des Calmouques sont le[s] plus grands de tous.*" Here then a race of animals, and one of the largest too, has been increased in its dimensions by *cold* and *moisture*, in direct opposition to the hypothesis, which supposes that these two circumstances diminish animal bulk, and that it is their contraries *heat* and *dryness* which enlarge it. But when we appeal to experience, we are not to rest satisfied with a single fact. Let us therefore try our question on more general ground. Let us take two portions of the earth, Europe and America for instance, sufficiently extensive to give operation to general causes; let us consider the circumstances peculiar to each, and observe their effect on animal nature. America, running through the torrid as well as the temperate zone, has more heat, collectively taken, than Europe. But Europe, according to our hypothesis, is the *dryest*. They are equally adapted then to animal productions; each being endowed with one of those causes which befriend animal growth, and with one which opposes it. If it be thought unequal to compare Europe with America, which is so much larger, I answer, not more so than to compare America with the whole world. Besides, the purpose of the comparison is to try an hypothesis, which makes the size of animals depend on the *heat* and *moisture* of climate. If therefore we take a region, so extensive as to comprehend a sensible distinction of climate, and so extensive too as that local accidents, or the intercourse of animals on its borders, may not materially affect the size of those in its interior parts, we shall comply with those conditions which the hypothesis may reasonably demand. The objection would be the weaker in the present case, because any intercourse of animals which may take place on the confines of Europe and Asia, is to the advantage of the former, Asia producing certainly larger animals than Europe. Let us then take a comparative view of the quadrupeds of Europe and America, presenting them to the eye in three different tables, in one of which shall be enumerated those found in both countries; in a second those found in one only; in a third those which have been domesticated in both. To facilitate the comparison, let those of each table be arranged in gradation according to their sizes, from the greatest to the smallest, so far as their sizes can be conjectured. The weights of the large animals shall be expressed in the English avoirdupoise pound and its decimals: those of the smaller in the ounce and its decimals. Those which are marked thus *, are actual weights of particular subjects, deemed among the largest of their species. Those marked thus †, are furnished by judicious persons, well acquainted with the species, and saying, from conjecture only, what the largest individual they had seen would probably have weighed. The other weights are taken from Messrs. Buffon and D'Aubenton, and are of such subjects as came casually to their hands for dissection. This circumstance must be remembered where their weights and mine stand opposed: the latter being stated, not to produce a conclusion in favour of the American species, but to justify a suspension of opinion until we are better informed, and a suspicion in the mean time that there is no uniform difference in favour of either; which is all I pretend.

I have not inserted in the first table the Phoca[3] nor leather-winged bat, because the one living half the year in the water, and the other being a winged animal, the individuals of each species may visit both continents.

Of the animals in the 1st table Mons. de Buffon himself informs us, [XXVII. 130. XXX. 213.] that the beaver, the otter, and shrew mouse, though of the same species, are larger in America than Europe. This should therefore have corrected the generality of his expressions XVIII. 145. and elsewhere, that the animals common to the two countries are considerably less in America than in Europe, "& cela sans ancune exception." He tells us too [Quadrup. VIII. 334. edit. Paris, 1777] that on examining a bear from America, he remarked no difference "dans le *forme* de cet ours d'Amerique comparé a celui d'Europe." But adds from Bartram's journal, that an American bear weighed 400 lb. English, equal to 367 lb. French: whereas we find the European bear examined by Mons. D'Aubenton [XVII. 82] weighed but 141 lb. French. That the palmated elk is larger in America than Europe we are informed by Kalm, a naturalist who visited the former by public appointment for the express purpose of examining the subjects of natural history. In this fact Pennant concurs with him. The same Kalm tells us[4] that the black moose, or renne of America, is as high as a tall horse; and Catesby,[5] that it is about the bigness of a middle sized ox. The same account of their size has been given me by many who have seen them. But Mons. D'Aubenton says[6] that the renne of Europe is about the size of a red deer.

A Comparative View of the Quadrupeds of Europe and of America.

I. ABORIGINALS OF BOTH.

	Europe. *lb.*	*America.* *lb.*
Mammoth		
Buffalo. Bison		*1800
White bear. Ours blanc		
Caribou. Renne		
Bear. Ours	153.7	*410
Elk. Elan. Original, palmated		
Red deer. Cerf	288.8	*273
Fallow deer. Daim	167.8	
Wolf. Loup	69.8	
Roe. Chevreuil	56.7	
Glutton. Gloüton. Carcajou		
Wild cat. Chat sauvage		†30
Lynx. Loup cervier	25.	
Beaver. Castor	18.5	*45
Badger. Blaireau	13.6	
Red fox. Renard	13.5	
Grey fox. Isatis		
Otter. Loutre	8.9	†12
Monax. Marmotte	6.5	
Vison. Fouine	2.8	
Hedgehog. Herisson	2.2	
Martin. Marte	1.9	†6
	oz.	
Water rat. Rat d'eau	7.5	
Wesel. Belette	2.2	*oz.*
Flying squirrel. Polatouche	2.2	†4
Shrew mouse. Musaraigne	1.	

II. ABORIGINALS OF ONE ONLY.

	Europe. *lb.*	*America.* *lb.*	
Sanglier. Wild boar	280.	Tapir	534.
Mouflon. Wild sheep	56.	Elk, round horned	†450
Bouquetin. Wild goat		Puma	
Lievre. Hare	7.6	Jaguar	218.
Lapin. Rabbit	3.4	Cabiai	109.
Putois. Polecat	3.3	Tamanoir	109.
Genette	3.1	Tamandua	65.4
Desman. Muskrat	*oz.*	Cougar of N. Amer.	75.
Ecureuil. Squirrel	12.	Cougar of S. Amer.	59.4
Hermine. Ermin	8.2	Ocelot	
Rat. Rat	7.5	Pecari	46.6
Loirs	3.1	Jaguaret	43.6
Lerot. Dormouse	1.8	Alco	
Taupe. Mole	1.2	Lama	

Continued

II. Continued

	Europe. lb.	America. lb.	
Hamster	.9	Paco	
Zisel		Paca	32.7
Leming		Serval	
Souris. Mouse	.6	Sloth. Unau	27¼
		Sanicovienne	
		Kincajou	
		Tatou Kabassou	21.8
		Urson. Urchin	
		Raccoon. Raton	16.5
		Coati	
		Coendou	16.3
		Sloth. Aï	13.
		Sapajou Ouarini	
		Sapajou Coaita	9.8
		Tatou Encubert	
		Tatou Apar	
		Tatou Cachica	7.
		Little Coendou	6.5
		Opossum. Sarigue	
		Tapeti	
		Margay	
		Crabier	
		Agouti	4.2
		Sapajou Saï	3.5
		Tatou Cirquinçon	
		Tatou Tatouate	3.3
		Mouffette Squash	3.3
		Mouffette Chinche	
		Mouffette Conepate. Scunk	
		Mouffette. Zorilla	
		Whabus. Hare. Rabbit	
		Aperea	
		Akouchi	
		Ondatra. Muskrat	
		Pilori	
		Great grey squirrel	†2.7
		Fox squirrel of Virginia	†2.625
		Surikate	2.
		Mink	†2.
		Sapajou. Sajou	1.8
		Indian pig. Cochon d'Inde	1.6
		Sapajou. Saïmiri	1.5
		Phalanger	
		Coquallin	
		Lesser grey squirrel	†1.5
		Black squirrel	†1.5
		Red squirrel	10 *oz.*
		Sagoin Saki	
		Sagoin Pinche	
		Sagoin Tamarin	*oz.*

Continued

II. Continued

	Europe. *lb.*	America. *lb.*	
		Sagoin Quistiti	4.4
		Sagoin Marikine	
		Sagoin Mico	
		Cayopollin	
		Fourmillier	
		Marmose	
		Sarigue of Cayenne	
		Tucan	
		Red mole	*oz.*
		Ground squirrel	4.

III. DOMESTICATED IN BOTH.

	Europe. *lb.*	America. *lb.*
Cow	763.	*2500
Horse		*1366
Ass		
Hog		*1200
Sheep		*125
Goat		*80
Dog	67.6	
Cat	7.	

The wesel is larger in America than in Europe, as may be seen by comparing its dimensions as reported by Mons. D'Aubenton[7] and Kalm. The latter tells us,[8] that the lynx, badger, red fox, and flying squirrel, are the *same* in America as in Europe: by which expression I understand, they are the same in all material circumstances, in size as well as others: for if they were smaller, they would differ from the European. Our grey fox is, by Catesby's account[9] little different in size and shape from the European fox. I presume he means the red fox of Europe, as does Kalm, where he says,[10] that in size "they do not quite come up to our foxes." For proceeding next to the red fox of America, he says "they are entirely the same with the European sort." Which shews he had in view one European sort only, which was the red. So that the result of their testimony is, that the American grey fox is somewhat less than the European red; which is equally true of the grey fox of Europe, as may be seen by comparing the measures of the Count de Buffon and Mons. D'Aubenton.[11] The white bear of America is as large as that of Europe. The bones of the mammoth which have been found in America, are as large as those found in the old world. It may be asked, why I insert the mammoth, as if it still existed? I ask in return, why I should omit it, as if it did not exist? Such is the œconomy of nature, that no instance can be produced of her having permitted any one race of her animals to become extinct; of her having formed any link in her great work so weak as to be broken. To add to this, the traditionary testimony of the Indians, that this animal still exists in the northern and western parts of America, would be adding the light of a taper to that of the meridian sun. Those parts still remain in their aboriginal state, unexplored and undisturbed by us, or by others for us. He may as well exist there now, as he did formerly where we find his bones. If he be a carnivorous animal, as some anatomists have conjectured, and the Indians affirm, his early retirement may be accounted for from the general destruction of the wild game by the Indians, which commences in the first instant of their connection with us, for the purpose of purchasing matchcoats, hatchets, and fire locks, with their skins. There remain then the buffalo, red deer, fallow deer, wolf, roe, glutton, wild cat, monax, vison, hedgehog, martin, and water rat, of the comparative sizes of which we have not sufficient testimony. It does not appear that Messrs. de Buffon and D'Aubenton have measured, weighed, or seen those of America. It is said of some of them, by some travellers, that they are smaller than the European. But who were these travellers? Have they not been men of a very different description from those who have laid open to us the other three quarters of the world? Was natural history the object of their travels? Did they measure or weigh the animals they speak of? or did they not judge of them by sight, or perhaps even from report only? Were they acquainted with the animals of their own country, with which they undertake to compare them? Have they not been so ignorant as often to mistake the species? A true answer to these questions would probably lighten their authority, so as to render it insufficient for the foundation of an hypothesis. How unripe we yet are, for an accurate comparison of the animals of the two countries, will appear from the work of Mons. de Buffon. The ideas we should have formed of the sizes of some animals, from the information he had received at his first publications concerning them, are very different from what his subsequent communications give us. And indeed his candour in this can never be too much praised. One sentence of his book must do him immortal honour. "*J'aime autant une personne qui me releve d'une erreur, qu'une autre qui m'apprend une verité, parce qu'en effet une erreur corrigée est une verité.*"[12] He seems to have thought the cabiai he first examined wanted little of its full growth. "*Il n'était pas encore tout à fait adulte.*" Yet he weighed but 46½ lb. and[13] he found afterward,[14] that these animals, when full grown, weigh 100 lb. He had supposed, from the examination of a jaguar,[15] said to be two years old, which weighed but 16 lb. 12 oz. that, when he should have acquired his full growth, he would not be larger than a middle sized dog. But a subsequent account[16] raises his weight to 200 lb. Further information will, doubtless, produce further corrections. The wonder is, not that there is yet something in this great work to correct, but that there is so little. The result of this view then is, that of 26 quadrupeds common to both countries, 7 are said to be larger in America, 7 of equal size, and 12 not sufficiently examined. So that the first table impeaches the first member of the assertion, that of the animals common to both countries, the American are smallest, "*et cela sans aucune exception.*" It shews it not just, in all the latitude in which its author has advanced it, and probably not to such a degree as to found a distinction between the two countries.

Proceeding to the second table, which arranges the animals found in one of the two countries only, Mons. de Buffon observes, that the tapir, the elephant of America, is but of the size of a small cow. To preserve our comparison, I will add that the wild boar, the elephant of Europe, is little more than half that size. I have made an elk with round or cylindrical horns, an animal of America, and peculiar to it; because I have seen many of them myself, and more of their horns; and because I can say, from the best information, that, in Virginia, this kind of elk has abounded much, and still exists in smaller numbers; and I could never learn that the palmated kind had been seen here at all. I suppose this confined to the more northern latitudes.[17] I have made our hare or rabbit peculiar, believing it to be different from both the European animals of those denominations, and calling it therefore by its Algonquin name Whabus, to keep it distinct from these. Kalm is of the same opinion.[18] I have enumerated the squirrels according to our own knowledge, derived from daily sight of them, because I am not able to reconcile with that the European appellations and descriptions. I have heard of other species, but they have never come within my own notice. These, I think, are the only instances in which I have departed from the authority of Mons. de Buffon in the construction of this table. I take him for my ground work, because I think him the best informed of any naturalist who has ever written. The result is, that there are 18 quadrupeds peculiar to Europe; more than four times as many, to wit 74, peculiar to America; that the first of these 74 weighs more than the whole column of Europeans;[19] and consequently this second table disproves the second member of the assertion, that the animals peculiar to the new world are on a smaller scale, so far as that assertion relied on European animals for support: and it is in full opposition to the theory which makes the animal volume to depend on the circumstances of *heat* and *moisture*.

The IIId. table comprehends those quadrupeds only which are domestic in both countries. That some of these, in some parts of America, have become less than their original stock, is doubtless true; and the reason is very obvious. In a thinly peopled country, the spontaneous productions of the forests and waste fields are sufficient to support indifferently the domestic animals of the farmer, with a very little aid from him in the severest and scarcest season. He therefore finds it more convenient to receive them from the hand of nature in that indifferent state, than to keep up their size by a care and nourishment which would cost him much labour. If, on this low fare, these animals dwindle, it is no more than they do in those parts of Europe where the poverty of the soil, or poverty of the owner, reduces them to the same scanty subsistance. It is the uniform effect of one and the same cause, whether acting on this or that side of the globe. It would be erring therefore against that rule of philosophy, which teaches us to ascribe like effects to like causes, should we impute this diminution of size in America to any imbecility or want of uniformity in the operations of nature. It may be affirmed with truth that, in those countries, and with those individuals of America, where necessity or curiosity has produced equal attention as in Europe to the nourishment of animals, the horses, cattle, sheep, and hogs of the one continent are as large as those of the other. There are particular instances, well attested, where individuals of this country have imported good breeders from England, and have improved their size by care in the course of some years. To make a fair comparison between the two countries, it will not answer to bring together animals of what might be deemed the middle or ordinary size of their species; because an error in judging of that middle or ordinary size would vary the result of the comparison. Thus Monsieur D'Aubenton[20] considers a horse of 4 feet 5 inches high and 400 lb. weight French, equal to 4 feet 8.6 inches and 436 lb. English, as a middle sized horse. Such a one is deemed a small horse in America. The extremes must therefore be resorted to. The same anatomist[21] dissected a horse of 5 feet 9 inches height, French measure, equal to 6 feet 1.7 English. This is near 6 inches higher than any horse I have seen: and could it be supposed that I had seen the largest horses in America, the conclusion would be, that ours have diminished, or that we have bred from a smaller stock. In Connecticut and Rhode-Island, where the climate is favourable to the production of grass, bullocks have been slaughtered which weighed 2500, 2200, and 2100 lb. nett; and those of 1800 lb. have been frequent. I have seen a hog weigh 1050 lb. after the blood, bowels, and hair had been taken from him.[22] Before he was killed an attempt was made to weigh him with a pair of steel-yards, graduated to 1200 lb. but he weighed more. Yet this hog was probably not within fifty generations of the European stock. I am well informed of another which weighed 1100 lb. gross. Asses have been still more neglected than any other domestic animal in America. They are neither fed nor housed in the most rigorous season of the year. Yet they are larger than those measured by Mons. D'Aubenton,[23] of 3 feet $7\frac{1}{4}$ inches, 3 feet 4 inches, and 3 feet $2\frac{1}{2}$ inches, the latter weighing only 215.8 lb. These sizes, I suppose, have been produced by the same negligence in Europe, which has produced a like diminution here. Where care has been taken of them on that side of the water, they have been raised to a size bordering on that of the horse; not by the *heat* and *dryness* of the climate, but by good food and shelter. Goats have been also much neglected in America. Yet they are very prolific here, bearing twice or three times a year, and from one to five kids at a birth. Mons. de Buffon has been sensible of a difference in this circumstance in favour of America.[24] But what are their greatest weights I cannot say. A large sheep here weights 100 lb. I observe Mons. D'Aubenton calls a ram of 62 lb. one of the middle size.[25] But to say what are the extremes of growth in these and the other domestic animals of America, would require information of which no one individual is possessed. The weights actually known and stated in the third table preceding will suffice to shew, that we may conclude, on probable grounds, that, with equal food and care, the climate of America will preserve the races of domestic animals as large as the European stock from which they are derived; and consequently that the third member of Mons. de Buffon's assertion, that the domestic animals are subject to degeneration from the climate of America, is as probably wrong as the first and second were certainly so.

That the last part of it is erroneous, which affirms the species of American quadrupeds are comparatively few, is evident from the tables taken altogether. By these it appears that there are an hundred species aboriginal of America. Mons. de Buffon supposes about double that number existing on the whole earth.[26] Of these Europe, Asia, and Africa, furnish suppose 126; that is, the 26 common to Europe and America, and about 100 which are not in America at all. The American species then are to those of the rest of the earth, as 100 to 126, or 4 to 5. But the residue of the earth being double the extent of America, the exact proportion would have been but as 4 to 8.

NOTES

1. XVII, 100–156.
2. VIII, 134.
3. It is said, that this animal is seldom seen above 30 miles from shore, or beyond the 56th degree of latitude. The interjacent islands between Asia and America admit his passage from one continent to the other without exceeding these bounds. And in fact, travellers tell us that these islands are places of principal resort for them, and specially in the season of bringing forth their young.
4. Ib. 233.
5. I, xxvil.
6. XXIV. 162.
7. XV. 42.
8. I. 359. I. 43, 221, 251. II. 52.
9. II. 78.
10. I. 220.
11. XXVII. 63. XLV. 119. Harris, II. 387. Buffon. Quad. IX.1.
12. Quad. IX. 158.
13. XXV. 184.
14. Quad. IX. 132.
15. XIX. 2.
16. Quad. IX. 41.
17. The descriptions of Theodat, Denys and La Hontan, cited by Mons. de Buffon under the article Elan, authorise the supposition, that the flat-horned elk is found in the northern parts of America. It has not however extended to our latitudes. On the other hand, I could never learn that the round-horned has been further North than the Hudson's river. This agrees with the former elk in its general character, being, like that when compared with a deer, very much larger, its ears longer broader, and thicker in proportion, its hair much longer, neck and tail shorter, having a dewlap before the breast (caruncula gutturalia Linnaei) a white spot often, if not always, a foot in diameter, on the hinder part of the buttocks round the tail; its gait a trot, and attended with a rattling of the hoofs; but distinguished from that decisively by its horns, which are not palmated, but round and pointed. This is the animal described by Catesby as the Cervus major Arnericanus, the stag of America, le cerf de l'Amérique. But it differs from the Cervus as totally, as does the palmated elk from the dama. And in fact it seems to stand in the same relations to the palmated elk, as the red deer does to the fallow. It has abounded in Virginia, has been seen, within my knowledge, on the eastern side of the Blue ridge since the year 1765, is now common beyond those

mountains, has been often brought to us and tamed, and their horns are in the hands of many. I should designate it as the "Alces Americanus cornibus teretibus." It were to be wished, that naturalists, who are acquainted with the renne and elk of Europe, and who may hereafter visit the northern parts of America, would examine well the animals called there by the names of grey and black moose, caribou, original, and elk. Mons. de Buffon has done what could be done from the materials in his hands, towards clearing up the confusion introduced by the loose application of these names among the animals they are meant to designate. He reduces the whole to the renne and flat-horned elk. From all the information I have been able to collect, I strongly suspect they will be found to cover three, if not four distinct species of animals. I have seen skins of a moose, and of the caribou: they differ more from each other, and from that of the round-horned elk, than I ever saw two skins differ which belonged to different individuals of any wild species. These differences are in the colour, length, and coarseness of the hair, and in the size, texture, and marks of the skin. Perhaps it will be found that there is, 1. The moose, black and grey, the former being said to be the male, and the latter the female. 2. The caribou or renne. 3. The flat-horned elk, or original. 4. The round-horned elk. Should this last, though possessing to nearly the characters of the elk, be found to be the same with the Cerf d'Ardennes or Brandhirtz of Germany, still I there will remain the three species first enumerated.

18. Kalm II. 340 I. 82.
19. The Tapir is the largest of the animals peculiar to America. I collect his weight thus. Mons. de Buffon says, XXIII. 274. that he is of the size of a Zebu, or a small cow. He gives us the measures of a Zebu, lb. 94, as taken by himself, viz. 5 feet 7 inches from the muzzle to the root of the tail, and 5 feet 1 inch circumference behind the fore legs. A bull, measuring in the same way 6 feet 9 inches and 5 feet 2 inches, weighed 600 lb. VIII. 153. The Zebu then, and of course the Tapir, would weigh about 500 lb. But one individual of every species of European peculiars would probably weigh less than 400 lb. These are French measures and weights.
20. VII. 432.
21. VII. 474.
22. In Williamsburg, April, 1769.
23. VIII. 48.55.66.
24. XVIII. 96.
25. IX. 41.
26. XXX. 219.

6

William Bartram (1739–1823)

As the son of naturalist John Bartram, William Bartram was remarkably well trained as a botanist, spending most of his childhood at the famed botanical gardens founded by his father on the Schuykill River near Philadelphia. Bartram had also accompanied his father on numerous botanizing expeditions throughout the American colonies during his youth. After a few disastrous business ventures, Bartram received financial support from a family friend and embarked on the journeys described in his Travels. *Bartram's descriptions of his trip throughout the southern colonies from 1773 to 1777 reflect not only a thorough training in natural history, but also display an enthusiasm for wild nature that is remarkable for his time. Additionally, his perceptive and sympathetic portrayals of the Indians he met on his journeys reflect his Quaker upbringing and his own philosophy of the merits of the "natural man."*

Bartram's Travels *was not published until 1791, and reaction to the book in America was tepid. In Europe, however, the response was far more favorable; Samuel Taylor Coleridge, for one, enthused that the* Travels *was a "series of poems . . . a* delicious *book." It is an irony of literary history that while American readers largely ignored Bartram's work, it was received enthusiastically by the European Romantics who in turn were such a large influence on the development of Transcendentalism—the movement that was so integral to the development of nature writing in the United States.*

From *Travels Through North and South Carolina, Georgia, East and West Florida, etc.* (1791)

After four days moderate and pleasant travelling, we arrived in the evening at the Buffalo Lick. This extraordinary place occupies several acres of ground, at the foot of the S. E. promontory of the Great Ridge, which, as before observed, divides the rivers Savanna and Alatamaha. A large cane swamp and meadows, forming an immense plain, lies S. E. from it; in this swamp I believe the head branches of the great Ogeeche river take their rise. The place called the Lick contains three or four acres, is nearly level, and lies between the head of the cane swamp and the ascent of the Ridge. The earth, from the superficies to an unknown depth, is an almost white or cinereous coloured tenacious fattish clay, which all kinds of cattle lick into great caves, pursuing the delicious vein. It is the common opinion of the inhabitants, that this clay is impregnated with saline vapours, arising from fossil salts deep in the earth; but I could discover nothing saline in its taste, but I imagined an insipid sweetness. Horned cattle, horses, and deer, are immoderately fond of it, insomuch, that their excrement, which almost totally covers the earth to some distance round this place, appears to be perfect clay; which, when dried by the sun and air, is almost as hard as brick.

We were detained at this place one day, in adjusting and planning the several branches of the survey. A circumstance occurred during this time, which was a remarkable instance of Indian sagacity, and had nearly disconcerted all our plans, and put an end to the business. The surveyor having fixed his compass on the staff, and about to ascertain the course from our place of departure, which was to strike Savanna river at the confluence of a certain river, about seventy miles distance from us; just as he had determined upon the point, the Indian Chief came up, and observing the course he had fixed upon, spoke, and said it was not right; but that the course to the place was so and so, holding up his hand, and pointing. The surveyor replied, that he himself was certainly right, adding, that that little instrument (pointing to the compass) told him so, which, he said, could not err. The Indian answered, he knew better, and that the little wicked instrument was a liar; and he would not acquiesce in its decisions, since it would wrong the Indians out of their land. This mistake (the surveyor proving to be in the wrong) displeased the Indians; the dispute arose to that height, that the Chief and his party had determined to break up the business, and return the shortest way home, and forbad the surveyors to proceed any farther: however, after some delay, the complaisance and prudent conduct of the Colonel made them change their resolution; the Chief became reconciled, upon condition that the compass should be discarded, and rendered incapable of serving on this business; that the Chief himself should lead the survey; and, moreover, receive an order for a very considerable quantity of goods.

Matters being now amicably settled, under this new regulation, the Colonel having detached two companies on separate routes, Mr. M'Intosh and myself attaching ourselves to the Colonel's party, whose excursion was likely to be the most extensive and varied, we set off from the Buffalo Lick, and the Indian Chief, heading the party, conducted us on a straight line, as appeared by collateral observation, to the desired place. We pursued nearly a north course up the Great Ridge, until we came near the branches of Broad River, when we turned off to the right hand, and encamped on a considerable branch of it. At this place we continued almost a whole day, constituting surveyors and astronomers, who were to take the course, distance, and observations on Broad River, and from thence down to its confluence with the Savanna.

The Great Ridge consists of a continued high forest; the soil fertile, and broken into moderately elevated hills, by the many rivulets which have their sources in it. The heights and precipices abound in rock and stone. The forest trees and other vegetable productions are the same as already mentioned about Little River: I observed Halesia, Styrax, Aesculus pavia, Aesc. sylvatica, Robinia hispida, Magnolia acuminata, Mag. tripetala, and some very curious new shrubs and plants, particularly the Physic-nut, or Indian Olive. The stems arise many from a root, two or three feet high; the leaves sit opposite, on every (sic) short petioles; they are broad, lanceolate, entire, and undulated, having smooth surfaces of a deep green colour. From the bosom of each leaf is produced a single oval drupe, standing erect, on long slender stems; it has a large kernel, and thin pulp. The fruit is yellow when ripe, and about the size of an olive. The Indians, when they go in pursuit of deer, carry this fruit with them, supposing that it has the power of charming or drawing that creature to them; from whence, with the traders, it has obtained the name of the Physic-nut, which means, with them, charming, conjuring, or fascinating. Malva scandens, Felix scandens, perhaps species of Trichomanes; the leaves are palmated, or radiated; it climbs and roves about, on shrubs, in moist ground. A very singular and elegant plant, of an unknown family, called Indian Lettuce, made its first appearance in these rich vales; it is a biennial; the primary or radical leaves are somewhat spatuled, or broad, lanceolate, and obtuse pointed, of a pale yellowish green, smooth surface, and of a delicate frame, or texture; these leaves, spread equally on every side, almost reclining on the ground; from their centre arises a strait upright stem, five, six or seven feet high, smooth and polished; the ground of a dark purple colour, which is elegantly powdered with greenish yellow specks; this stem, three fourths of its length, is embellished with narrow leaves, nearly of the same form of the radical ones, placed at regular distances, in verticillate order. The superior one fourth division of this stem is formed into a pyramidal spike of flowers, rather diffuse; these flowers are of the hexandria, large, and expanded; of a dark purple colour, delicately powdered with green, yellow, and red, and divided into six parts, or petals; these are succeeded by triquetrous dry pericarps, when ripe.

This great ridge is a vast extended projection of the Cherokee or Alegany mountains, gradually encreasing in height and extent, from its extremity at the Lick, to its union with the high ridge of mountains anciently called the Apalachian mountains; it every where approaches much nearer the waters of the Alatamaha than those of the Savanna: at one particular place, where we encamped, on the Great Ridge, during our repose there, part of a day. Our hunters going out, and, understanding that their route was to the low lands on the Ocone, I accompanied them: we had not rode above three miles before we came to the banks of that beautiful river. The cane swamps, of immense extent, and the oak forests, on the level lands, are incredibly fertile; which appears from the tall reeds of the one, and the heavy timber of the other.

Before we left the waters of Broad River, having encamped in the evening, on one of its considerable branches, and left my companions, to retire, as usual, on botanical researches, on ascending a steep rocky hill, I accidentally discovered a new species of Caryophyllata (Geum odoratissimum) on reaching to a shrub, my foot slipped, and, in recovering myself, I tore up some of the plants, whose roots filled the air with animating scents of cloves and spicy perfumes.

On my return towards camp, I met my philosophic companion, Mr. M'Intosh, who was seated on the bank of a rivulet, and whom I found highly entertained by a very novel and curious natural exhibition, in which I participated with high relish. The waters at this place were still and shoal, and flowed over a bed of gravel just beneath a rocky rapid: in this eddy shoal were a number of little gravelly pyramidal hills, whose summits rose almost to the surface of the water, very artfully constructed by a species of small cray-fish (Cancer macrourus) which inhabited them: here seemed to be their citadel, or place of retreat for their young, against the attacks and ravages of their enemy, the gold-fish: these, in numerous bands, continually infested them, except at short intervals, when small detachments of veteran cray-fish sallied out upon them, from their cells within the gravelly pyramids, at which time a brilliant sight presented: the little gold-fish instantly fled from every side, darting through the transparent waters like streams of lightning; some even sprang above the surface, into the air, but all quickly returned to the charge, surrounding the pyramids as before, on the retreat of the cray-fish; in this manner the war seemed to be continual.

The gold-fish is about the size of the anchovy, nearly four inches long, of a neat slender form; the head is covered with a salade of an ultramarine blue, the back of a reddish brown, the sides and belly of a flame, or of the colour of a fine red lead; a narrow dusky line runs along each side, from the gills to the tail; the eyes are large, with the iris like burnished gold. This branch of Broad River is about twelve yards wide, and has two, three, and four feet depth of water, and winds through a fertile vale, almost overshadowed on one side by a ridge of high hills, well timbered with Oak, Hiccory, Liriodendron, Magnolia acuminata, Pavia sylvatica, and on their rocky summits, Fagus castania, Rhododendron ferruginium, Kalmia latifolia, Cornus Florida, &c.

One of our Indian young men, this evening, caught a very large salmon trout, weighing about fifteen pounds, which he presented to the Col. who ordered it to be served up for supper. The Indian struck this fish, with a reed harpoon, pointed very sharp, barbed, and hardened by the fire. The fish lay close under the steep bank, which the Indian discovered and struck with his reed; instantly the fish darted off with it, whilst the Indian pursued, without extracting the harpoon, and with repeated thrusts drowned it, and then dragged it to shore. After leaving Broad River, the land rises very sensibly, and the country being mountainous, our progress became daily more difficult and slow; yet the varied scenes of pyramidal hills, high forests, rich vales, serpentine rivers, and cataracts, fully compensated for our difficulties and delays. I observed the great Aconitum napellus, Delphinium perigrinum, the carminativc Angelica lucida,[1] and cerulean Malva.

We at length happily accomplished our line, arriving at the little river, where our hunters bringing in plenty of venison and turkeys, we had a plentiful feast at supper. Next morning we marked the corner tree, at the confluence of Little river and the Savanna; and, soon after, the Indians amicably took leave of us, returning home to their towns.

The rocks and fossils, which constitute the hills of this middle region, are of various species, as, Quartsum, Ferrum, Cos, Silex, Glarea, Arena, Ochra, Stalectites, Saxum, Mica, &c. I saw no signs of Marble, Plaster, or Lime-stone; yet there is, near Augusta, in the forests, great piles of a porous friable white rock, in large and nearly horizontal masses, which seems to be an heterogeneous concrete, consisting of pulverized sea shells, with a small proportion of sand; it is soft, and easily wrought into any form, yet of sufficient consistence for constructing any building.

As for the animal productions, they are the same which originally inhabited this part of North America, except such as have been affrighted away since the invasion of the Europeans. The buffalo (Urus) once so very numerous, is not at this day to be seen in this part of the country; a few elk, and those only in the Apalachian mountains. The dreaded and formidable rattlesnake is yet too common, and a variety of other serpents abound, particularly that admirable creature the glass-snake: I saw a very large and beautiful one, a little distance from our camp. The allegator, a species of crocodile, abounds in the rivers and swamps, near the sea coast, but is not to be seen above Augusta. Bears, tygers,[2] wolves, and wild cats (Felis cauda truncata) are numerous enough; and there is a very great variety of Papilio and Phalina, many of which are admirably beautiful, as well as other insects of infinite variety.

The surveyors having completed their observations, we set off next day on our return to Augusta, taking our route generally through the low lands on the banks of the Savanna. We crossed Broad River, at a newly settled plantation, near its confluence with the Savanna, On my arrival at Augusta, finding myself a little fatigued, I staid there a day or two, and then set off again for Savanna, the capital, where we arrived in good health.

Having, in this journey, met with extraordinary success, not only in the enjoyment of an uninterrupted state of good health, and escaping ill accidents, incident to such excursions, through uninhabited wildernesses, and an Indian frontier, but also in making a very extensive collection of new discoveries of natural productions. On the recollection of so many and great favours and blessings, I now, with a high sense of gratitude, presume to offer up my sincere thanks to the Almighty, the Creator and Preserver.

NOTES

1. Called Nondo in Virginia: by the Creek and Cherokee traders, White Root.
2. This creature is called, in Pennsylvania and the northern States, Panther; but in Carolina and the southern States, is called Tyger; they are very strong, much larger than any dog, of a yellowish brown, or clay colour, having a very long tail; they are a mischievous animal, and prey on calves, young colts, &c.

7

Red Jacket (Segoyewatha) (ca. 1750–1830)

Like most of the Six Nations of the Iroquois Confederacy, the Seneca nation sided with the British during the American Revolution, a choice that had disastrous consequences for that western New York nation. Red Jacket served primarily as a messenger for the British forces, and had a reputation as a warrior that was, at best, mixed. He was not considered a significant leader until after the war, when his talent as an orator became apparent. Red Jacket's oratory at a series of treaty councils between the Americans and the Seneca nation, held from the 1780s to the 1820s, became legendary, however, his eloquence did little to stop the cession of the Seneca nation's traditional territory to federal, state, and private interests. Red Jacket did support the United States in the War of 1812, and distinguished himself in battle. For the remainder of his life, he continued to resist white inroads on Indian culture and beliefs and the spread of Christianity among his people, although his influence was apparently diminished by his own problems with alcohol. He died in January 1830 at the Buffalo Creek Reservation in New York.

Red Jacket's reply to Reverend Cram, a member of the Boston Missionary Society seeking to convert the Seneca people to Christianity, was first published in 1809 and was reprinted widely thereafter. In his speech, Red Jacket counters the missionary's entreaties to convert by pointing out that in the past the whites had called the Indians their brothers while continuing to take their lands until "we have scarcely a place left to spread our blankets. You have got our country, but are not satisfied; you want to force your religion upon us." This contemporary transcript of Red Jacket's speech is therefore of interest both as an early record of a Native American's reaction to territorial concessions, and as a document of resistance to the work of Christian missionaries among Native peoples.

Red Jacket's Reply to Reverend Cram (1809)

Indian Speech.

[In the summer of 1805, a number of the principal Chiefs and Warriors of the Six Nations of Indians, principally Senecas, assembled at Buffalo Creek, in the State of New-York, at the particular request of a gentleman Missionary from the State of Massachusetts.[1] The Missionary being furnished with an Interpreter, and accompanied by the Agent of the United States for Indian affairs, met the Indians in Council, when the following talk took place.]

First, by the Agent

"Brothers of the Six Nations; I rejoice to meet you at this time, and thank the Great Spirit, that he has preserved you in health, and given me another opportunity of taking you by the hand.

"Brothers; The person who sits by me, is a friend who has come a great distance to hold a talk with you. He will inform you what his business is, and it is my request that you would listen with attention to his words."

MISSIONARY. "My Friends; I am thankful for the Opportunity afforded us of uniting together at this time. I had a great desire to see you, and inquire into your state and welfare; for this purpose I have travelled a great distance, being sent by your old friends, the Boston Missionary Society. You will recollect they formerly sent missionaries among you, to instruct you in religion, and labour for your good. Although they have not heard from you for a long time, yet they have not forgotten their brothers the Six Nations, and are still anxious to do you good.

"Brothers: I have not come to get your lands or your money, but to enlighten your minds, and to instruct you how to worship the Great Spirit agreeably to his mind and will, and to preach to you the gospel of his son Jesus Christ. There is but one true religion, and but one way to serve God, and if you do not embrace the right way, you cannot be happy hereafter. You have never worshipped the Great Spirit in a manner acceptable to him; but have all your lives been in great errours and darkness. To endeavour to remove these errours, and open your eyes, so that you might see clearly, is my business with you.

"Brothers: I wish to talk with you as one friend talks with another; and if you have any objections to receive the religion which I preach, I wish you to state them; and I will endeavour to satisfy your minds, and remove the objections.

"Brothers: I want you to speak your minds freely; for I wish to reason with you on the subject, and, if possible, remove all doubts, if there be any on your minds. The subject is an important one, and it is of consequence that you

First published in the *Monthly Anthology and Boston Review*, April 1809, Vol. 6, pp. 221–224.

give it an early attention while the offer is made you. Your friends, the Boston Missionary Society, will continue to send you good and faithful ministers, to instruct and strengthen you in religion, if, on your part, you are willing to receive them.

"Brothers; Since I have been in this part of the country, I have visited some of your small villages, and talked with your people. They appear willing to receive instruction, but, as they look up to you as their older brothers in council, they first want to know your opinion on the subject.

"You have now heard what I have to propose at present. I hope you will take it into consideration, and give me an answer before we part."

After about two hours consultation among themselves, the Chief commonly called by the white people, Red Jacket,[2] rose and spoke as follows:

"Friend and Brother; It was the will of the Great Spirit that we should meet together this day. He orders all things, and has given us a fine day for our Council. He has taken his garment from before the sun, and caused it to shine with brightness upon us. Our eyes are opened, that we see clearly; our ears are unstopped, that we have been able to hear distinctly the words you have spoken. For all these favors we thank the Great Spirit; and Him only.

"Brother; This council fire was kindled by you. It was at your request that we came together at this time. We have listened with attention to what you have said. You requested us to speak our minds freely. This gives us great joy; for we now consider that we stand upright before you, and can speak what we think. All have heard your voice, and all speak to you now as one man. Our minds have agreed.

"Brother; You say you want an answer to your talk before you leave this place. It is right you should have one, as you are a great distance from home, and we do not wish to detain you. But we will first look back a little, and tell you what our fathers have told us, and what we have heard from the white people.

"Brother; Listen to what we say.

"There was a time when our forefathers owned this great island. Their seats extended from the rising to the setting sun. The Great Spirit had made it for the use of Indians. He had created the buffalo, the deer, and other animals for food. He had made the bear and the beaver. Their skins served us for clothing. He had scattered them over the country, and taught us how to take them. He had caused the earth to produce corn for bread. All this He had done for his red children, because He loved them. If we had some disputes about our hunting ground, they were generally settled without the shedding of much blood. But an evil day came upon us. Your forefathers crossed the great water, and landed on this island. Their numbers were small. They found friends and not enemies. They told us they had fled from their own country for fear of wicked men, and had come here to enjoy their religion. They asked for a small seat. We took pity on them, granted their request; and they sat down amongst us. We gave them corn and meat, they gave us poison [alluding, it is supposed, to ardent spirits] in return.

"The white people had now found our country. Tidings were carried back, and more came amongst us. Yet we did not fear them. We took them to be friends. They called us brothers. We believed them, and gave them a larger seat. At length their numbers had greatly increased. They wanted more land; they wanted our country. Our eyes were opened, and our minds became uneasy. Wars took place. Indians were hired to fight against Indians, and many of our people were destroyed. They also brought strong liquor amongst us. It was Strong and powerful, and has slain thousands.

"Brother; Our seats were once large and yours were small. You have now become a great people, and we have scarcely a place left to spread our blankets. You have got our country, but are not satisfied; you want to force your religion upon us.

"Brother; Continue to listen.

"You say that you are sent to instruct us how to worship the Great Spirit agreeably to his mind, and, if we do not take hold of the religion which you white people teach, we shall be unhappy hereafter. You say that you are right and we are lost. How do we know this to be true? We understand that your religion is written in a book. If it was intended for us as well as you, why has not the Great Spirit given to us, and not only to us, but why did he not give to our forefathers the knowledge of that book, with the means of understanding it rightly? We only know what you tell us about it. How shall we know when to believe, being so often deceived by the white people?

"Brother; You say there is but one way to worship and serve the Great Spirit. If there is but one religion; why do you white people differ so much about it? Why not all agreed, as you can all read the book?

"Brother; We do not understand these things.

"We are told that your religion was given to your forefathers and has been handed down from father to son. We also have a religion, which was given to our forefathers, and has been handed down to us their children. We worship in that way. It teaches us to be thankful for all the favors we receive; to love each other, and to be united. We never quarrel about religion.

"Brother; The Great Spirit has made us all, but he has made a great difference between his white and red children. He has given us different complexions and different customs. To you He has given the arts. To these He has not opened our eyes. We know these things to be true. Since He has made so great a difference between us in other things; why may we not conclude that He has given us a different religion according to our understanding? The Great Spirit does right. He knows what is best for his children; we are satisfied.

"Brother; We do not wish to destroy your religion or take it from you. We only want to enjoy our own.

"Brother; We are told that you have been preaching to white people in this place. These people are our neighbors. We are acquainted with them. We will wait a little while, and see what effect your preaching has upon them. If we find it does them good, makes them honest, and less disposed to cheat Indians; we will then consider again of what you have said.

"Brother; You have now heard our answer to your talk, and this is all we have to say at present.

"As we are going to part, we will come and take you by the hand, and hope the Great Spirit will protect you on your journey, and return you safe to your friends."

As the Indians began to approach the missionary, he rose hastily from his seat and replied, that he could not take them by the hand; that there was no fellowship between the religion of God and the works of the devil.

This being interpreted to the Indians, they smiled, and retired in a peaceable manner.

It being afterwards suggested to the missionary that his reply to the Indians was rather indiscreet; he observed, that he supposed the ceremony of shaking hands would be received by them as a token that he assented to what was said. Being otherwise informed, he said he was very sorry for the expressions.

NOTES

1. Reverend Cram.
2. His Indian name is Sagu-yu-what-ha; which interpreted is, Keeper awake.

8

Ralph Waldo Emerson (1803–1882)

Emerson was the son of a Unitarian minister who died while Emerson was still a boy. After graduating from Harvard in 1821, it appeared that he would follow in his father's footsteps when he enrolled as a student at the Divinity School at Harvard and was later admitted as a candidate to the ministry. In 1832, however, following the death of his first wife, Emerson resigned from his pastorate at the Second Church of Boston and left for a tour of Europe where he met Carlyle, Coleridge, and Wordsworth. Emerson and his second wife, Lydia, relocated to Concord, Massachusetts where he wrote his first book, Nature *(1836), which became a manifesto for the philosophical movement known as Transcendentalism. Emerson soon established a reputation as one of America's leading writers and philosophers and made the small village of Concord an intellectual center of the United States. As Walt Whitman wrote in a letter to Emerson that was included in the 1856 edition of* Leaves of Grass, *"You have discovered that new moral American continent without which . . . the physical continent remained incomplete . . . It is yours to have been the original true Captain who put to sea, intuitive, rendering the first report."*

In 1837 Emerson delivered his address "The American Scholar" to the Phi Beta Kappa Society at Harvard. Like Nature, *"The American Scholar" is crucial to understanding Emerson's transcendentalist philosophy; like* Nature, *it has also been read as an intellectual declaration of independence from Europe. While Emerson was not as interested in the facts of natural history as was his friend Henry Thoreau (as John Burroughs wrote, Emerson went to the woods primarily "to fetch the word of the wood-god to men"), nature study was an integral part of his philosophy. In "The American Scholar," Emerson posited that the student of nature "shall see that nature is the opposite of the soul, answering to it part for part. One is seal and one is print. Its beauty is the beauty of his own mind. Its laws are the laws of his own mind. Nature then becomes to him the measure of his attainments. So much of nature as he is ignorant of, so much of his own mind does he not yet possess. And, in fine, the ancient precept, 'Know thyself,' and the modern precept, 'Study nature,' become at last one maxim." Over the next several decades Emerson wrote and lectured extensively, often on this link between God, nature, and the individual soul. It is nearly impossible to overestimate the influence of Emerson on the development of American nature writing; writers such as Henry Thoreau, John Burroughs, and John Muir, among many others, all cite Emerson as a primary influence on their own writing. In the last of many essays that Burroughs wrote about the philosopher during the course of his own lengthy literary career, Burroughs asserted that Emerson remained such an important figure that "attention cannot be directed to him too often."*

"The American Scholar" (1837)

Mr. President and Gentlemen:

I greet you on the recommencement of our literary year. Our anniversary is one of hope, and, perhaps, not enough of labor. We do not meet for games of strength or skill, for the recitation of histories, tragedies, and odes, like the ancient Greeks; for parliaments of love and poesy, like the Troubadours; nor for the advancement of science, like our contemporaries in the British and European capitals. Thus far, our holiday has been simply a friendly sign of the survival of the love of letters amongst a people too busy to give to letters any more. As such it is precious as the sign of an indestructible instinct. Perhaps the time is already come when it ought to be, and will be, something else; when the sluggard intellect of this continent will look from under its iron lids and fill the postponed expectation of the world with something better than the exertions of mechanical skill. Our day of dependence, our long apprenticeship to the learning of other lands, draws to a close. The millions that around us are rushing into life, cannot always be fed on the sere remains of foreign harvests. Events, actions arise, that must be sung, that will sing themselves. Who can doubt that poetry will revive and lead in a new age, as the star in the constellation Harp, which now flames in our zenith, astronomers announce, shall one day be the polestar for a thousand years?

In this hope I accept the topic which not only usage but the nature of our association seem to prescribe to this day—the AMERICAN SCHOLAR. Year by year we come up hither to read one more chapter of his biography. Let us inquire what light new days and events have thrown on his character and his hopes.

It is one of those fables which out of an unknown antiquity convey an unlooked-for wisdom, that the gods, in the beginning, divided Man into men, that he might be more helpful to himself: just as the hand was divided into fingers, the better to answer its end.

The old fable covers a doctrine ever new and sublime; that there is One Man—present to all particular men only partially, or through one faculty; and that you must take the whole society to find the whole man. Man is not a farmer, or a professor, or an engineer, but he is all. Man is priest, and scholar, and statesman, and producer, and soldier. In the *divided* or social state these functions are parcelled out to individuals, each of whom aims to do his stint of the joint work, whilst each other performs his. The fable implies that the individual, to possess himself, must sometimes return from his own labor to embrace all the other laborers. But, unfortunately, this original unit, this fountain of power, has been so distributed to multitudes, has been so minutely subdivided and peddled out, that it is spilled into drops, and cannot be gathered. The state of society is one in which the members have suffered amputation from the trunk, and strut about so many walking monsters—a good finger, a neck, a stomach, an elbow, but never a man.

Man is thus metamorphosed into a thing, into many things. The planter, who is Man sent out into the field to gather food, is seldom cheered by any idea of the true dignity of his ministry. He sees his bushel and his cart, and nothing beyond, and sinks into the farmer, instead of Man on the farm. The tradesman scarcely ever gives an ideal worth to his work, but is ridden by the routine of his craft, and the soul is subject to dollars. The priest becomes a form; the attorney a statute-book; the mechanic a machine; the sailor a rope of the ship.

In this distribution of functions the scholar is the delegated intellect. In the right state he is *Man Thinking*. In the degenerate state, when the victim of society, he tends to become a mere thinker, or still worse, the parrot of other men's thinking.

In this view of him, as Man Thinking, the theory of his office is contained. Him Nature solicits with all her placid, all her monitory pictures; him the past instructs; him the future invites. Is not indeed every man a student, and do not all things exist for the student's behoof? And, finally, is not the true scholar the only true master? But the old oracle said, "All things have two handles: beware of the wrong one." In life, too often, the scholar errs with mankind and forfeits his privilege. Let us see him in his school, and consider him in reference to the main influences he receives.

I. The first in time and the first in importance of the influences upon the mind is that of nature. Every day, the sun; and, after sunset, Night and her stars. Ever the winds blow; ever the grass grows. Every day, men and women, conversing—beholding and beholden. The scholar is he of all men whom this spectacle most engages. He must settle its value in his mind. What is nature to him? There is never a beginning, there is never an end, to the inexplicable continuity of this web of God, but always circular power returning into itself. Therein it resembles his own spirit, whose beginning, whose ending, he never can find—so entire, so boundless. Far too as her splendors shine, system on system shooting like rays, upward, downward, without centre, without circumference—in the mass and in the particle, Nature hastens to render account of herself to the mind. Classification begins. To the young mind every thing is individual, stands by itself. By and by, it finds how to join two things and see in them one nature; then three, then three thousand; and so, tyrannized over by its own unifying instinct, it goes on tying things together, diminishing anomalies, discovering roots running under ground whereby contrary and remote things cohere and flower out from one stem. It presently learns that since the dawn of history there has been a constant accumulation and classifying of facts. But what is classification but the perceiving that these objects are not chaotic, and are not foreign, but have a law which is also a law of the human mind? The astronomer discovers that geometry, a pure abstraction of the human mind, is the measure of planetary motion. The chemist finds proportions and intelligible method throughout matter; and science is nothing but the finding of analogy, identity, in the most remote parts. The ambitious soul sits down before each refractory fact; one after another reduces all strange constitutions, all new powers, to their class and their law, and goes on forever to animate the last fibre of organization, the outskirts of nature, by insight.

Thus to him, to this schoolboy under the bending dome of day, is suggested that he and it proceed from one root; one is leaf and one is flower; relation, sympathy, stirring in every vein. And what is that root? Is not that the soul of his soul? A thought too bold; a dream too wild. Yet when this spiritual light shall have revealed the law of more earthly natures—when he has learned to worship the soul, and to see that the natural philosophy that now is, is only the first gropings of its gigantic hand, he shall look forward to an ever expanding knowledge as to a becoming creator. He shall see that nature is the opposite of the soul, answering to it part for part. One is seal and one is print. Its beauty is the beauty of his own mind. Its laws are the laws of his own mind. Nature then becomes to him the measure of his attainments. So much of nature as he is ignorant of, so much of his own mind does he not yet possess. And, in fine, the ancient precept, "Know thyself," and the modern precept, "Study nature," become at last one maxim.

* * *

It is a mischievous notion that we are come late into nature; that the world was finished a long time ago. As the world was plastic and fluid in the hands of God, so it is ever to so much of his attributes as we bring to it. To ignorance and sin, it is flint. They adapt themselves to it as they may; but in proportion as a man has any thing in him divine, the firmament flows before him and takes his signet and form. Not he is great who can alter matter, but he who can alter my state of mind. They are the kings of the world who give the color of their present thought to all nature and all art, and persuade men by the cheerful serenity of their carrying the matter, that this thing which they do is the apple which the ages have desired to pluck, now at last ripe, and inviting nations to the harvest. The great man makes the great thing. Wherever Macdonald sits, there is the head of the table. Linnæus makes botany the most alluring of studies, and wins it from the farmer and the herbwoman; Davy, chemistry; and Cuvier, fossils. The day is always his who works in it with serenity and great aims. The unstable estimates of men crowd to him whose mind is filled with a truth, as the heaped waves of the Atlantic follow the moon.

9

George Catlin (1796–1872)

After a brief career as a lawyer, George Catlin decided to pursue his lifelong interest in painting and set up a portrait studio in Philadelphia in the 1820s. After painting the portrait of the explorer William Clark, coleader of the Lewis and Clark Expedition, Catlin accompanied Clark (who now served as Superintendent for Indian Affairs) on a trip west to negotiate treaties with several of the Indian nations. Catlin became fascinated with the Indians and their way of life, which he realized would soon be forever changed as a result of Western expansion. In 1832 he embarked on an eight-year-long painting expedition that took him through much of the Indian territories in North America. Catlin visited approximately 146 Indian tribes during his journeys and painted and sketched thousands of portraits and other scenes of Indian life. Upon his return east, Catlin had hoped to sell his collection of paintings to the nation, but when that proposal was defeated in the U.S. Senate, he left for Europe, where he spent the next thirty years with his family. He returned to the United States shortly before his death in 1872.

In addition to his paintings and sketches, Catlin wrote an illustrated book of his journeys, Letters and Notes of the Manners, Customs, and Condition of the North American Indians, *which was published in 1841. In* Letters and Notes *he described a way of life that he already realized was soon to disappear. In the passage included here, he recounts a remarkable "reverie" in which he foresees the disappearance of both the buffalo and the Indian tribes who depended upon that animal for their sustenance. During his expedition, Catlin also sent a number of dispatches back to newspapers in the east, and in one of those reports he made the earliest known calls for the creation of a system of national parks, a proposal that was reiterated in* Letters and Notes.

From *Letters and Notes of the Manners, Customs, and Condition of the North American Indians* (1841)

LETTER—NO. 31.

Mouth of Teton River, *Upper Missouri.*

* * *

It is truly a melancholy contemplation for the traveller in this country, to anticipate the period which is not far distant, when the last of these noble animals [the bison], at the hands of white and red men, will fall victims to their cruel and improvident rapacity; leaving these beautiful green fields, a vast and idle waste, unstocked and unpeopled for ages to come, until the bones of the one and the traditions of the other will have vanished, and left scarce an intelligible trace behind.

That the reader should not think me visionary in these contemplations, or romancing in making such assertions, I will hand him the following item of the extravagancies which are practiced in these regions, and rapidly leading to the results which I have just named.

When I first arrived at this place, on my way up the river, which was in the month of May, in 1832, and had taken up my lodgings in the Fur Company's Fort, Mr. Laidlaw, of whom I have before spoken, and also his chief clerk, Mr. Halsey, and many of their men, as well as the chiefs of the Sioux, told me, that only a few days before I arrived, (when an immense herd of buffaloes had showed themselves on the opposite side of the river, almost blackening the plains for a great distance,) a party of five or six hundred Sioux Indians on horseback, forded the river about mid-day, and spending a few hours amongst them, recrossed the river at sun-down and came into the Fort with *fourteen hundred fresh buffalo tongues*, which were thrown down in a mass, and for which they required but a few gallons of whiskey, which was soon demolished, indulging them in a little, and harmless carouse.

This profligate waste of the lives of these noble and useful animals, when, from all that I could learn, not a skin or a pound of the meat (except the tongues), was brought in, fully supports me in the seemingly extravagant predictions that I have made as to their extinction, which I am certain is near at hand. In the above extravagant instance, at a season when their skins were without fur and not worth taking off, and their camp was so well stocked with fresh and dried meat, that they had no occasion for using the flesh, there is a fair exhibition of the improvident character of the savage, and also of his recklessness in catering for his appetite, so long as the present inducements are held out to him in his country, for its gratification.

In this singular country, where the poor Indians have no laws or regulations of society, making it a vice or an impropriety to drink to excess, they think it no harm to indulge in the delicious beverage, as long as they are able to buy whiskey to drink. They look to white men as wiser than themselves, and able to set them examples—they see none of these in their country but sellers of whiskey, who are constantly tendering it to them, and most of them setting the example by using it themselves; and they easily acquire a taste, that to be catered for, where whiskey is sold at sixteen dollars per gallon, soon impoverishes them, and must soon strip the skin from the last buffalo's back that lives in their country, to "be dressed by their squaws" and vended to the Traders for a pint of diluted alcohol.

From the above remarks it will be seen, that not only the red men, but red men and white, have aimed destruction at the race of these animals; and with them, *beasts* have turned hunters of buffaloes in this country, slaying them, however, in less numbers, and for far more laudable purpose than that of selling their skins. The white wolves, of which I have spoken in a former epistle, follow the herds of buffaloes as I have said, from one season to another, glutting themselves on the carcasses of those that fall by the deadly shafts of their enemies, or linger with disease or old age to be dispatched by these sneaking cormorants, who are ready at all times kindly to relieve them from the pangs of a lingering death.

Whilst the herd is together, the wolves never attack them, as they instantly gather for combined resistance, which they effectually make. But when the herds are travelling, it often happens that an aged or wounded one, lingers at a distance behind, and when fairly out of sight of the herd, is set upon by these voracious hunters, which often gather to the number of fifty or more, and are sure at last to torture him to death, and use him up at a meal. The buffalo, however, is a huge and furious animal, and when his retreat is cut off, makes desperate and deadly resistance, contending to the last moment for the right of life—and oftentimes deals death by wholesale, to his canine assailants, which he is tossing into the air or stamping to death under his feet.

During my travels in these regions, I have several times come across such a gang of these animals surrounding arm old or a wounded bull, where it would seem, from appearances, that they had been for several days in attendance, and at intervals desperately engaged in the effort to take his life. But a short time since, as one of my hunting companions and myself were returning to our encampment with our horses loaded with meat, we discovered at a distance, a huge bull, encircled with a gang of white wolves; we rode up as near as we could without driving them away, and being within pistol shot, we had a remarkably good view, where I sat for a few moments and made a sketch in my note-book; after which, we rode up and gave the signal for them to disperse, which they instantly did, withdrawing themselves to the distance of fifty or sixty rods, when we found, to our great surprise, that the animal had made desperate resistance, until his eyes were entirely eaten out of his head—the grizzle of his nose was mostly gone—his tongue was half eaten off, and the skin and flesh of his legs torn almost literally into strings. In this tattered and torn condition, the poor old veteran stood bracing up in the midst of his devourers, who had ceased hostilities for a few minutes, to enjoy a sort of parley, recovering strength and preparing to resume the attack in a few moments again. In this group, some were reclining, to gain breath, whilst others were sneaking about and licking their chaps in anxiety for a renewal of the attack; and others, less lucky, had been crushed to death by the feet or the horns of the bull. I rode nearer to the pitiable object as he stood bleeding and trembling before me, and said to him, "Now is your time, old fellow, and you had better be off." Though blind and nearly destroyed, there seemed evidently to be a recognition of a friend in me, as he straightened up, and, trembling with excitement, dashed off at full speed upon the prairie, in a straight line. We turned our horses and resumed our march, and when we had advanced a mile or more, we looked back, and on our left, where we saw again the ill-fated animal surrounded by his tormentors, to whose insatiable voracity he unquestionably soon fell a victim.

Thus much I wrote of the buffaloes, and of the accidents that befall them, as well as of the fate that awaits them; and before I closed my book, I strolled out one day to the shade of a plum-tree, where I laid in the grass on a favourite bluff, and wrote thus:—

"It is generally supposed, and familiarly said, that a man *'falls'* into a rêverie; but I seated myself in the shade a few minutes since, resolved to *force* myself into one; and for this purpose I laid open a small pocket-map of North America, and excluding my thoughts from every other object in the world, I soon succeeded in producing the desired illusion. This little chart, over which I bent, was seen in all its parts, as nothing but the green and vivid reality. I was lifted up upon an imaginary pair of wings, which easily raised and held me floating in the open air, from whence I could behold beneath me the Pacific and the Atlantic Oceans—the great cities of the East, and the mighty rivers. I could see the blue chain of the great lakes at the North—the Rocky Mountains, and beneath them and near their base, the vast, and almost boundless plains of grass, which were speckled with the bands of grazing buffaloes!

"The world turned gently around, and I examined its surface; continent after continent passed under my eye, and yet amidst them all, I saw not the vast and vivid green, that is spread like a carpet over the Western wilds of my own country. I saw not elsewhere in the world, the myriad herds of buffaloes—my eyes scanned in vain, for they were not. And when I turned again to the wilds of my native land, I beheld them all in motion! For the distance of several hundreds of miles from North to South, they were wheeling about in vast columns and herds—some were scattered, and ran with furious wildness—some lay dead, and others were pawing the earth for a hiding-place—some were sinking down and dying, gushing out their life's blood in deep-drawn sighs—and others were contending in furious battle for the life they possessed, and the ground that they stood upon. They had long since assembled from the thickets, and secret haunts of the deep forest, into the midst of the treeless and bushless plains, as the place for their safety. I could see in an hundred places, amid the wheeling bands, and on their skirts and flanks, the leaping wild horse darting among them. I saw not the arrows, nor heard the twang of the sinewy bows that sent them; but I saw their victims fall!—on other steeds that rushed along their sides, I saw the glistening lances, which seemed to lay across them; their blades were blazing in the sun, till dipped in blood, and then I lost them! In other parts (and there were many), the vivid flash of *fire-arms* was seen—*their* victims fell too, and over their dead bodies hung suspended in air, little clouds of whitened smoke, from under which the flying horsemen had darted forward to mingle again with, and deal death to, the trampling throng.

"So strange were men mixed (both red and white) with the countless herds that wheeled and eddyed about, that all below seemed one vast extended field of battle—whole armies, in some places, seemed to blacken the earth's surface;—in other parts, regiments, battalions, wings, platoons, rank and file, and '*Indian-file*'—all were in motion; and death and destruction seemed to be the watch-word amongst them. In their turmoil, they sent up great clouds of dust, and with them came the mingled din of groans and trampling hoofs, that seemed like the rumbling of a dreadful cataract, or the roaring of distant thunder. Alternate pity and admiration harrowed up in my bosom and my brain, many a hidden thought; and amongst them a few of the beautiful notes that were once sung, and exactly in point: '*Quadrupedante putrem sonitu qu tit ungula campum.*' Even such was the din amidst the quadrupeds of these vast plains. And from the craggy cliffs of the Rocky Mountains also were seen descending into the valley, the myriad Tartars, who had not horses to ride, but before their well-drawn bows the fattest of the herds were falling. Hundreds and thousands were strewed upon the plains—they were flayed, and their reddened carcasses left; and about them bands of wolves, and dogs, and buzzards were seen devouring them. Contiguous, and in sight, were the distant and feeble smokes of wigwams and villages, where the skins were dragged, and dressed for white man's luxury! where they were all sold for *whiskey*, and the poor Indians laid drunk, and were crying. I cast my eyes into the towns and cities of the East, and there I beheld buffalo robes hanging at almost every door for traffic; and I saw also the curling smokes of a thousand *Stills*—and I said, 'Oh insatiable man, is thy avarice such! wouldst thou tear the skin from the back of the last animal of this noble race, *and rob thy fellow-man of his meat, and for it give him poison!*' "

* * *

Many are the rudenesses and wilds in Nature's works, which are destined to fall before the deadly axe and desolating hands of cultivating man; and so amongst her ranks of *living*, of beast and human, we often find noble stamps, or beautiful colours, to which our admiration clings; and even in the overwhelming march of civilized improvements and refinements do we love to cherish their existence, and lend our efforts to preserve them in their primitive rudeness. Such of Nature's works are always worthy of our preservation and protection; and the further we become separated (and the face of the country) from that pristine wildness and beauty, the more pleasure does the mind of enlightened man feel in recurring to those scenes, when he can have them preserved for his eyes and his mind to dwell upon.

Of such "rudenesses and wilds," Nature has no where presented more beautiful and lovely scenes, than those of the vast prairies of the West; and of *man* and *beast*, no nobler specimens than those who inhabit them—the *Indian* and the *buffalo—joint* and original tenants of the soil, and fugitives together from the approach of civilized man; they have fled to the great plains of the West, and there, under an equal doom, they have taken up their *last abode*, where their race will expire, and their bones will bleach together.

It may be that *power* is *right*, and *voracity* a *virtue;* and that these people, and these noble animals, are *righteously* doomed to an issue that *will* not be averted. It can be easily proved—we have a civilized science that can easily do it, or anything else that may be required to cover the iniquities of civilized man in catering for his unholy appetites. It can be proved that the weak and ignorant have no *rights*—that there can be no virtue in darkness—that God's gifts have no meaning or merit until they are appropriated by civilized man—by him brought into the light, and converted to his use and luxury. We have a mode of reasoning (I forget what it is called) by which all this can be proved, and even more. The *word* and the *system* are entirely of *civilized* origin; and latitude is admirably given to them in proportion to the increase of civilized wants, which often require a *judge* to overrule the laws of nature. I say that *we* can prove such things; but an *Indian* cannot. It is a mode of reasoning unknown to him in his nature's simplicity, but admirably adapted to subserve the interests of the enlightened world, who are always their own judges, when dealing with the savage; and who, in the present refined age, have many appetites that can only be lawfully indulged, by proving God's laws defective.

It is not enough in this polished and extravagant age, that we get from the Indian his lands, and the very clothes from his back, but the food from their mouths must be stopped, to add a new and useless article to the fashionable world's luxuries. The ranks must be thinned, and the race exterminated, of this noble animal, and the Indians of the great plains left without the means of supporting life, that white men may figure a few years longer, enveloped in buffalo robes—that they may spread them, for their pleasure and elegance, over the backs of their sleighs, and trail them ostentatiously amidst the busy throng, as a thing of beauty and elegance that had been made for them!

Reader! listen to the following calculations, and forget them not. The buffaloes (the quadrupeds from whose backs your beautiful robes were taken, and whose myriads were once spread over the whole country, from the Rocky Mountains to the Atlantic Ocean) have recently fled before the appalling appearance of civilized man, and taken up their abode and pasturage amid the almost boundless prairies of the West. An instinctive dread of their deadly foes, who made an easy prey of them whilst grazing in the forest, has led them to seek the midst of the vast and treeless plains of grass, as the spot where they would be least exposed to the assaults of their enemies; and it is exclusively in those desolate fields of silence (yet of beauty) that they are to be found—and over these vast steppes, or prairies, have they fled, like the Indian, towards the "setting sun;" until their bands have been crowded together, and their limits confined to a narrow strip of country on this side of the Rocky Mountains.

This strip of country, which extends from the province of Mexico to lake Winnepeg on the North, is almost one entire plain of grass, which is, and ever must be, useless to cultivating man. It is here, and here chiefly, that the buffaloes dwell; and with, and hovering about them, live and flourish the tribes of Indians, whom God made for the enjoyment of that fair land and its luxuries.

It is a melancholy contemplation for one who has travelled as I have, through these realms, and seen this noble animal in all its pride and glory, to contemplate it so rapidly wasting from the world, drawing the irresistible conclusion too, which one must do, that its species is soon to be extinguished, and with it the peace and happiness (if not the actual existence) of the tribes of Indians who are joint tenants with them, in the occupancy of these vast and idle plains.

And what a splendid contemplation too, when one (who has travelled these realms, and can duly appreciate them) imagines them as they *might* in future be seen, (by some great protecting policy of government) preserved in their pristine beauty and wildness, in a *magnificent park*, where the world could see for ages to come, the native Indian in his classic attire, galloping his wild horse, with sinewy bow, and shield and lance, amid the fleeting herds of elks and buffaloes. What a beautiful and thrilling specimen for America to preserve and hold up to the view of her refined citizens and the world, in future ages! A *nation's Park*, containing man and beast, in all the wild and freshness of their nature's beauty!

I would ask no other monument to my memory, nor any other enrolment of my name amongst the famous dead, than the reputation of having been the founder of such an institution.

Such scenes might easily have been preserved, and still could be cherished on the great plains of the West, without detriment to the country or its borders; for the tracts of country on which the buffaloes have assembled, are uniformly sterile, and of no available use to cultivating man.

It is on these plains, which are stocked with buffaloes, that the finest specimens of the Indian race are to be seen. It is here, that the savage is decorated in the richest costume. It is here, and here only, that his wants are all satisfied, and even the *luxuries* of life are afforded him in abundance. And here also is he the proud and honourable man (before he has had teachers or laws), above the imported wants, which beget meanness and vice; stimulated by ideas of honour and virtue, in which the God of Nature has certainly not curtailed him.

There are, by a fair calculation, more than 300,000 Indians, who are now subsisted on the flesh of the buffaloes, and by those animals supplied with all the luxuries of life which they desire, as they know of none others. The great variety of uses to which they convert the body and other parts of that animal, are almost incredible to the person who has not actually dwelt amongst these people, and closely studied their modes and customs. Every part of their flesh is converted into food, in one shape or another, and on it they entirely subsist. The robes of the animals are worn by the Indians instead of blankets—their skins when tanned, are used as coverings for their lodges, and for their beds; undressed, they are used for constructing canoes—for saddles, for bridles—l'arrêts, lasos, and thongs. The horns are shaped into ladles and spoons—the brains are used for dressing the skins—their bones are used for saddle trees—for war clubs, and scrapers for graining the robes—and others are broken up for the marrow-fat which is contained in them. Their sinews are used for strings and backs to their bows—for thread to string their beads and sew their dresses. The feet of the animals are boiled, with their hoofs, for the glue they contain, for fastening their arrow points, and many other uses. The hair from the head and shoulders, which is long, is twisted and braided into halters, and the tail is used for a fly brush. In this wise do these people convert and use the various parts of this useful animal, and with all these luxuries of life about them, and their numerous games, they are happy (God bless them) in the ignorance of the disastrous fate that awaits them.

Yet this interesting community, with its sports, its wildnesses, its languages, and all its manners and customs, could be perpetuated, and also the buffaloes, whose numbers would increase and supply them with food for ages and centuries to come, if a system of non-intercourse could be established and preserved. But such is not to be the case—the buffalo's doom is sealed, and with their extinction must assuredly sink into real despair and starvation, the inhabitants of these vast plains, which afford for the Indians, no other possible means of subsistence; and they must at last fall a prey to wolves and buzzards, who will have no other bones to pick.

It seems hard and cruel, (does it not?) that we civilized people with all the luxuries and comforts of the world about us, should be drawing from the backs of these useful animals the skins for our luxury, leaving their carcasses to be devoured by the wolves—that we should draw from that country, some 150 or 200,000 of their robes annually, the greater part of which are taken from animals that are killed expressly for the robe, at a season when the meat is not cured and preserved, and for each of which skins the Indian has received but a pint of whiskey!

Such is the fact, and that number or near it, are annually destroyed, in addition to the number that is necessarily killed for the subsistence of 300,000 Indians, who live entirely upon them. It may be said, perhaps, that the Fur Trade of these great western realms, which is now limited chiefly to the purchase of buffalo robes, is of great and national importance, and should and must be encouraged. To such a suggestion I would reply, by merely enquiring, (independently of the poor Indians' disasters,) how much more advantageously would such a capital be employed, both for the weal of the country and for the owners, if it were invested in machines for the manufacture of *woollen robes*, of equal and superior value and beauty; thereby encouraging the growers of wool, and the industrious manufacturer, rather than cultivating a taste for the use of buffalo skins; which is just to be acquired, and then, from necessity, to be dispensed with, when a few years shall have destroyed the last of the animals producing them.

It may be answered, perhaps, that the necessaries of life are given in exchange for these robes; but what, I would ask, are the necessities in Indian life, where they have buffaloes in abundance to live on? The Indians' necessities are entirely artificial—are all created; and when the buffaloes shall have disappeared in his country, which will be within *eight* or *ten* years, I would ask, who is to supply him with the necessaries of life then? and I would ask, further, (and leave the question to be answered ten years hence), when the skin shall have been stripped from the back of the last animal, who is to resist the ravages of 300,000 starving savages; and in their trains, 1,500,000 wolves, whom direst necessity will have driven from their desolate and gameless plains, to seek for the means of subsistence along our exposed frontier? God has everywhere supplied man in a state of Nature, with the necessaries of life, and before we destroy the game of his country, or teach him new desires, he has no wants that are not satisfied.

Amongst the tribes who have been impoverished and repeatedly removed, the necessaries of life are extended with a better grace from the hands of civilized man; 90,000 of such have already been removed, and they draw from Government some 5 or 600,000 dollars annually in cash; *which money passes immediately into the hands of white men*, and for it the necessaries of life *may be* abundantly furnished. But who, I would ask, are to furnish the Indians who have been instructed in this unnatural mode—living upon *such* necessaries, and even luxuries of life, extended to them by the hands of white men, when those annuities are at an end, and the skin is stripped from the last of the animals which God gave them for their subsistence?

Reader, I will stop here, lest you might forget to answer these important queries—these are questions which I know will puzzle the world—and, perhaps, it is not right that I should ask them.

* * *

Thus much I wrote and painted at this place, whilst on my way up the river: after which I embarked on the steamer for the Yellow Stone, and the sources of the Missouri, through which interesting regions I have made a successful Tour; and have returned, as will have been seen by the foregoing narrations, in my canoe, to this place, from whence I am to descend the river still further in a few days. If I ever get time, I may give further Notes on this place, and of people and their doings, which I met with here; but at present, I throw my note-book, and canvass, and brushes into my canoe, which will be launched to-morrow morning, and on its way towards St. Louis, with myself at the steering-oar, as usual; and with Ba'tiste and Bogard to paddle, of whom, I beg the readers' pardon for having said nothing of late, though they have been my constant companions. Our way is now over the foaming and muddy waters of the Missouri, and amid snags and drift logs (for there is a sweeping freshet on her waters), and many a day will pass before other Letters will come from me; and possibly, the reader may have to look to my biographer for the rest. Adieu.

10

Henry David Thoreau (1817–1862)

While many of his contemporaries saw Henry Thoreau as little more than one of Ralph Waldo Emerson's Transcendentalist disciples, many scholars now trace the roots of modern environmentalism back to Thoreau. After Thoreau returned to his native village of Concord, Massachusetts following his graduation from Harvard in 1837, Ralph Waldo Emerson became his spiritual and literary mentor. Unlike Emerson, however, Thoreau's studies of natural history were far more systematic, and resulted in a body of work—particularly that contained in his remarkable journals—that anticipates the science of ecology. Thoreau's intimate knowledge of natural history distinguishes his work from that of Emerson in both content and style. After reading Thoreau's journals in 1863, Emerson wrote in his own: "I find the same thought, the same spirit that is in me, but he takes a step beyond, & illustrates by excellent images that which I should have conveyed in sleepy generalities."

On July 4, 1845, Thoreau moved into a small hut that he had built for himself on a piece of property owned by Emerson on Walden Pond, two miles outside the village of Concord. While living at Walden, Thoreau drafted his first book, A Week on the Concord and Merrimack Rivers *(1849) and began the book that would become the first classic work of American nature writing,* Walden *(1854). While living at Walden Pond, Thoreau was also arrested and spent a night in jail for refusing to pay his poll tax. Thoreau considered this tax a way for the government to prosecute the Mexican war, which many abolitionists (like Thoreau) saw as an attempt by slaveholding states to expand their territory. That experience was recounted in his great essay on civil disobedience, where Thoreau outlined the practice of passive resistance that would influence later civil rights leaders, including Mohandas Gandhi and Martin Luther King, Jr.*

Although Thoreau traveled relatively little (proudly proclaiming that "I have traveled a good deal in Concord") he did take three rather lengthy trips to visit the Maine woods, which were described in three essays—"Ktaadn," "Chesuncook," and "The Allegash and East Branch"—the first two of which were published in major periodicals during his lifetime. All three essays were gathered by his friend William Ellery Channing and published after Thoreau's death as The Maine Woods *(1864). Thoreau's spirituality was nature-based and he frequently criticized both organized religion and anthropocentrism. When "Chesuncook" was published in* The Atlantic Monthly *in 1858, the editor of that magazine, James Russell Lowell, excised (without Thoreau's permission) a passage that he found to be too close to paganism for his taste: "It is the living spirit of the tree, not its spirit of turpentine, with which I sympathize, and which heals my cuts. It is as immortal as I am, and perchance will go to as high a heaven, there to tower above me still." The essay as published in* The Atlantic *did, however contain Thoreau's call for the establishment of national forest preserves—one of the earliest such proposals to be made in a major American periodical.*

From "Chesuncook" (1858)

Humboldt has written an interesting chapter on the primitive forest, but no one has yet described for me the difference between that wild forest which once occupied our oldest townships, and the tame one which I find there to-day. It is a difference which would be worth attending to. The civilized man not only clears the land permanently to a great extent, and cultivates open fields, but he tames and cultivates to a certain extent, the forest itself. By his mere presence, almost, he changes the nature of the trees as no other creature does. The sun and air, and perhaps fire, have been introduced, and grain raised where it stands. It has lost its wild, damp, and shaggy look, the countless fallen and decaying trees are gone, and consequently that thick coat of moss which lived on them is gone too. The earth is comparatively bare and smooth and dry. The most primitive places left with us are the swamps, where the spruce still grows shaggy with usnea. The surface of the ground in the Maine woods is everywhere spongy and saturated with moisture. I noticed that the plants which cover the forest floor there are such as are commonly confined to swamps with us,—the *Clintonia borealis*, orchises, creeping snowberry, and others; and the prevailing aster there is the *Aster acuminatus*, which with us grows in damp and shady woods. The asters *cordifolius* and *macrophyllus* also are common, asters of little or no color, and sometimes without petals. I saw no soft, spreading, second-growth white-pines, with smooth bark, acknowledging the presence of the wood-chopper, but even the young white-pines were all tall and slender rough-barked trees.

Those Maine woods differ essentially from ours. There you are never reminded that the wilderness which you are threading is, after all, some villager's familiar wood-lot, some widow's thirds, from which her ancestors have sledded fuel for generations, minutely described in some old deed which is recorded, of which the owner has got a plan, too, and old bound-marks may be found every forty rods, if you will search. 'Tis true, the map may inform you that you stand on land granted by the State to some academy, or on Bingham's purchase; but these names do not impose on you, for you see nothing to remind you of the academy or of Bingham. What were the "forests" of England to these? One writer relates of the Isle of Wight, that in Charles the Second's time "there were woods in the island so complete and extensive, that it is said a squirrel might have traveled in several parts many leagues together on the top of the trees." If it were not for the rivers (and he might go round their heads), a squirrel could here travel thus the whole breadth of the country.

We have as yet had no adequate account of a primitive pine forest. I have noticed that in a physical atlas lately published in Massachusetts, and used in our schools, the "wood land" of North America is limited almost solely to the valleys of the Ohio and some of the Great Lakes, and the great pine forests of the globe are not represented. In our vicinity, for instance, New Brunswick and Maine are exhibited as bare as Greenland. It may be that the children of Greenville, at the foot of Moosehead Lake, who surely are not likely to be scared by an owl, are referred to the valley of the Ohio to get an idea of a forest; but they would not know what to do with their moose, bear, caribou, beaver, etc., there. Shall we leave it to an Englishman to inform us, that "in North America, both in the United States and Canada, are the most extensive pine forests in the world"? The greater part of New Brunswick, the northern half of Maine, and adjacent parts of Canada, not to mention the northeastern part of New York and other tracts farther off, are still covered with an almost unbroken pine forest.

But Maine, perhaps, will soon be where Massachusetts is. A good part of her territory is already as bare and commonplace as much of our neighborhood, and her villages generally are not so well shaded as ours. We seem to think that the earth must go through the ordeal of sheep-pasturage before it is habitable by man. Consider Nahant, the resort of all the fashion of Boston,—which peninsula I saw but indistinctly in the twilight, when I steamed by it, and thought that it was unchanged since the discovery. John Smith described it in 1614 as "the Mattahunts, two pleasant isles of groves, gardens, and cornfields;" and others tell us that it was once well wooded, and even furnished timber to build the wharves of Boston. Now it is difficult to make a tree grow there, and the visitor comes away with a vision of Mr. Tudor's ugly fences, a rod high, designed to protect a few pear-shrubs. And what are we coming to in our Middlesex towns?—a bald, staring town-house, or meeting-house, and a bare liberty-pole, as leafless as it is fruitless, for all I can see. We shall be obliged to import the timber for the last, hereafter, or splice such sticks as we have;—and our ideas of liberty are equally mean with these. The very willow-rows lopped every three years for fuel or powder,—and every sizable pine and oak, or other forest tree, cut down within the memory of man! As if individual speculators were to be allowed to export the clouds out of the sky, or the stars out of the firmament, one by one. We shall be reduced to gnaw the very crust of the earth for nutriment.

They have even descended to smaller game. They have lately, as I hear, invented a machine for chopping up huckleberry-bushes fine, and so converting them into fuel!—bushes which, for fruit alone, are worth all the pear-trees in the country many times over. (I can give you a list of the three best kinds, if you want it.) At this rate, we shall all be obliged to let our beards grow at least, if only to hide the nakedness of the land and make a sylvan appearance. The farmer sometimes talks of "brushing up," simply as if bare ground looked better than clothed ground, than that which wears its natural vesture,—as if the wild hedges, which, perhaps, are more to his children than his whole farm beside, were *dirt.* I know of one who deserves to be called the Tree-hater, and, perhaps, to leave this for a new patronymic to his children. You would think that he had been warned by an oracle that he would be killed by the fall of a tree, and so was resolved to anticipate them. The journalists think that they cannot say too much in favor of such "improvements" in husbandry; it is a safe theme, like piety; but as for the beauty of one of these "model farms," I would as lief see a patent churn and a man turning it. They are, commonly, places merely where somebody is making money, it may be counterfeiting. The virtue of making two blades of grass grow where only one grew before does not begin to be superhuman.

Nevertheless, it was a relief to get back to our smooth, but still varied landscape. For a permanent residence, it seemed to me that there could be no comparison between this and the wilderness, necessary as the latter is for a resource and a background, the raw material of all our civilization. The wilderness is simple, almost to barrenness. The partially cultivated country it is which chiefly has inspired, and will continue to inspire, the strains of poets, such as compose the mass of any literature. Our woods are sylvan, and their inhabitants woodmen and rustics,—that is, *selvaggia*, and the inhabitants are *salvages.* A civilized man, using the word in the ordinary sense, with his ideas and associations, must at length pine there, like a cultivated plant, which clasps its fibres about a crude and undissolved mass of peat. At the extreme North, the voyagers are obliged to dance and act plays for employment. Perhaps our own woods and fields,—in the best wooded towns, where we need not quarrel about the huckleberries,—with the primitive swamps scattered here and there in their midst, but not prevailing over them, are the perfection of parks and groves, gardens, arbors, paths, vistas, and landscapes. They are the natural consequence of what art and refinement we as a people have,—the common which each village possesses, its true paradise, in comparison with which all elaborately and willfully wealth-constructed parks and gardens are paltry imitations. Or, I would rather say, such *were* our groves twenty years ago. The poet's, commonly, is not a logger's path, but a woodman's. The logger and pioneer have preceded him, like John the Baptist; eaten the wild honey, it may be, but the locusts also; banished decaying wood and the spongy mosses which feed on it, and built hearths and humanized Nature for him.

But there are spirits of a yet more liberal culture, to whom no simplicity is barren. There are not only stately pines, but fragile flowers, like the orchises, commonly described as too delicate for cultivation, which derive their nutriment from the crudest mass of peat. These remind us, that, not only for strength, but for beauty, the poet must, from time to time, travel the logger's path and the Indian's trail, to drink at some new and more bracing fountain of the Muses, far in the recesses of the wilderness.

The kings of England formerly had their forests "to hold the king's game," for sport or food, sometimes destroying villages to create or extend them; and I think that they were impelled by a true instinct. Why should not we, who have renounced the king's authority, have our national preserves, where no villages need be destroyed, in which the bear and panther, and some even of the hunter race, may still exist, and not be "civilized off the face of the earth,"—our forests, not to hold the king's game merely, but to hold and preserve the king himself also, the lord of creation,—not for idle sport or food, but for inspiration and our own true recreation? or shall we, like the villains, grub them all up, poaching on our own national domains?

11

George Perkins Marsh (1801–1882)

By the early 1800s, most of southern Vermont had been deforested, the land cleared for farms and pastures. While still a boy, Marsh observed the effects of such deforestation firsthand when a flood destroyed his father's sawmill. Like his father, Marsh trained as an attorney, and opened a law office in Burlington, Vermont in 1825. He was elected to Congress twice (1844–1848), where he was instrumental in the creation of the Smithsonian Institution. After losing his congressional seat, Marsh had a lengthy diplomatic career that included appointments as ambassador to Turkey and a twenty-year term as ambassador in Italy.

While living in Italy Marsh completed work on Man and Nature *(1864), a work he had originally intended to call "Man the Disturber of Nature's Harmonies." The publication of* Man and Nature *is a landmark event in American environmental history, a work that the eminent critic Lewis Mumford has described as "the fountainhead of the conservation movement." As was the case with Henry Thoreau, Marsh anticipated the modern science of ecology, and his book was the first to provide extensive rationales for watershed and forest protection. Marsh's prose reflects his legal training, for better and for worse; his arguments are carefully constructed and supported heavily with evidence, but his writing style is often ponderous, with footnotes of daunting length. Still, his ecological insight is remarkable, particularly for a layperson, as is his prescient views on such topics as the need for reforestation and on the untapped potential of solar power. Marsh's work was cited in support of conservation measures for over fifty years, and provided important supporting evidence for proponents of the creation of the Adirondack and Catskill forest preserves in New York State.*

From *Man and Nature* (1864)

Destructiveness of Man.

Man has too long forgotten that the earth was given to him for usufruct alone, not for consumption, still less for profligate waste. Nature has provided

The text is from the 1874 edition, *The Earth as Modified by Human Action: A New Edition of Man and Nature.*

against the absolute destruction of any of her elementary matter, the raw material of her works; the thunderbolt and the tornado, the most convulsive throes of even the volcano and the earthquake, being only phenomena of decomposition and recomposition. But she has left it within the power of man irreparably to derange the combinations of inorganic matter and of organic life, which through the night of æons she had been proportioning and balancing, to prepare the earth for his habitation, when in the fulness of time his Creator should call him forth to enter into its possession.

Apart from the hostile influence of man, the organic and the inorganic world are, as I have remarked, bound together by such mutual relations and adaptations as secure, if not the absolute permanence and equilibrium of both, a long continuance of the established conditions of each at any given time and place, or at least, a very slow and gradual succession of changes in those conditions. But man is everywhere a disturbing agent. Wherever he plants his foot, the harmonies of nature are turned to discords. The proportions and accommodations which insured the stability of existing arrangements are overthrown. Indigenous vegetable and animal species are extirpated, and supplanted by others of foreign origin, spontaneous production is forbidden or restricted, and the face of the earth is either laid bare or covered with a new and reluctant growth of vegetable forms, and with alien tribes of animal life. These intentional changes and substitutions constitute, indeed, great revolutions; but vast as is their magnitude and importance, they are, as we shall see, insignificant in comparison with the contingent and unsought results which have flowed from them.

The fact that, of all organic beings, man alone is to be regarded as essentially a destructive power, and that he wields energies to resist which Nature—that nature whom all material life and all inorganic substance obey—is wholly impotent, tends to prove that, though living in physical nature, he is not of her, that he is of more exalted parentage, and belongs to a higher order of existences, than those which are born of her womb and live in blind submission to her dictates.

There are, indeed, brute destroyers, beasts and birds and insects of prey—all animal life feeds upon, and, of course, destroys other life,—but this destruction is balanced by compensations. It is, in fact, the very means by which the existence of one tribe of animals or of vegetables is secured against being smothered by the encroachments of another; and the reproductive powers of species, which serve as the food of others, are always proportioned to the demand they are destined to supply. Man pursues his victims with reckless destructiveness; and, while the sacrifice of life by the lower animals is limited by the cravings of appetite, he unsparingly persecutes, even to extirpation. thousands of organic forms which he cannot consume.[1]

The earth was not, in its natural condition, completely adapted to the use of man, but only to the sustenance of wild animals and wild vegetation. These live, multiply their kind in just proportion, and attain their perfect measure of strength and beauty, without producing or requiring any important change in the natural arrangements of surface, or in each other's spontaneous tendencies, except such mutual repression of excessive increase as may prevent the extirpation of one species by the encroachments of another. In short, without man, lower animal and spontaneous vegetable life would have been practically constant in type, distribution, and proportion, and the physical geography of the earth would have remained undisturbed for indefinite periods, and been subject to revolution only from slow development, from possible, unknown cosmical causes, or from geological action.

But man, the domestic animals that serve him, the field and garden plants the products of which supply him with food and clothing cannot subsist and rise to the full development of their higher properties, unless brute and unconscious nature be effectually combated, and, in a great degree, vanquished by human art. Hence, a certain measure of transformation of terrestrial surface, of suppression of natural, and stimulation of artificially modified productivity becomes necessary. This measure man has unfortunately exceeded. He has felled the forests whose network of fibrous roots bound the mould to the rocky skeleton of the earth; but had he allowed here and there a belt of woodland to reproduce itself by spontaneous propagation, most of the mischiefs which his reckless destruction of the natural protection of the soil has occasioned would have been averted. He has broken up the mountain reservoirs, the percolation of whose waters through unseen channels supplied the fountains that refreshed his cattle and fertilized his fields; but he has neglected to maintain the cisterns and the canals of irrigation which a wise antiquity had constructed to neutralize the consequences of its own imprudence. While he has torn the thin glebe which confined the light earth of extensive plains, and has destroyed the fringe of semi-aquatic plants which skirted the coast and checked the drifting of the sea sand, he has failed to prevent the spreading of the dunes by clothing them with artificially propagated vegetation. He has ruthlessly warred on all the tribes of animated nature whose spoil he could convert to his own uses, and he has not protected the birds which prey on the insects most destructive to his own harvests.

Purely untutored humanity, it is true, interferes comparatively little with the arrangements of nature,[2] and the destructive agency of man becomes more and more energetic and unsparing as he advances in civilization, until the impoverishment, with which his exhaustion of the natural resources of the soil is threatening him, at last awakens him to the necessity of preserving what is left, if not of restoring what has been wantonly wasted. The wandering savage grows no cultivated vegetable, fells no forest, and extirpates no useful plant, no noxious weed. If his skill in the chase enables him to entrap numbers of the animals on which he feeds, he compensates this loss by destroying also the lion, the tiger, the wolf, the otter, the seal, and the eagle, thus indirectly protecting the feebler quadrupeds and fish and fowls, which would otherwise become the booty of beasts and birds of prey. But with stationary life, or at latest with the pastoral state, man at once commences an almost indiscriminate warfare upon all the forms of animal and vegetable existence around him, and as be advances in civilization, he gradually eradicates or transforms every spontaneous product of the soil he occupies.[3]

Human and Brute Action Compared.

It is maintained by authorities as high as any known to modern science, that the action of man upon nature, though greater in *degree*, does not differ in *kind* from that of wild animals. It is perhaps impossible to establish a radical distinction *in genere* between the two classes of effects, but there is an essential difference between the motive of action which calls out the energies of civilized man and the mere appetite which controls the life of the beast. The action of man, indeed, is frequently followed by unforeseen and undesired results, yet it is nevertheless guided by a self-conscious will aiming as often at secondary and remote as at immediate objects. The wild animal, on the other hand, acts instinctively, and, so far as we are able to perceive, always with a view to single and direct purposes. The backwoodsman and the beaver alike fell trees; the man that he may convert the forest into an olive grove that will mature its fruit only for a succeeding generation, the beaver that he may feed upon the bark of the trees or use them in the construction of his habitation. The action of brutes upon the material world is slow and gradual, and usually limited, in any given case, to a narrow extent of territory. Nature is allowed time and opportunity to set her restorative powers at work, and the destructive animal has hardly retired from the field of his ravages before nature has repaired the damages occasioned by his operations. In fact, he is expelled from the scene by the very efforts which she makes for the restoration of her dominion. Man, on the contrary, extends his action over vast spaces, his revolutions are swift and radical, and his devastations are, for an almost incalculable time after he has withdrawn the arm that gave the blow, irreparable.

The form of geographical surface, and very probably the climate of a given country, depend much on the character of the vegetable life belonging to it. Man has, by domestication, greatly changed the habits and properties of the plants he rears; he has, by voluntary selection, immensely modified the forms and qualities of the animated creatures that serve him; and he has, at the same time, completely rooted out many forms of animal if not of vegetable being.[4] What is there, in the influence of brute life, that corresponds to this? We have no reason to believe that, in that portion of the American continent which, though peopled by many tribes of quadruped and fowl, remained uninhabited by man or only thinly occupied by purely savage tribes, any sensible geographical change had occurred within twenty centuries before the epoch of discovery and colonization, while, during the same period, man had changed millions of square miles, in the fairest and most fertile regions of the Old World, into the barrenest deserts.

The ravages committed by man subvert the relations and destroy the balance which nature had established between her organized and her inorganic creations, and she avenges herself upon the intruder, by letting loose upon her defaced provinces destructive energies hitherto kept in check by organic forces destined to be his best auxiliaries, but which he has unwisely dispersed and driven from the field of action. When the forest is gone, the great reservoir of moisture stored up in its vegetable mould is evaporated, and returns only in deluges of rain to wash away the parched dust into which that mould has been converted. The well-wooded and humid hills are turned to ridges of dry rock, which encumbers the low grounds and chokes the watercourses with its débris, and—except in countries favored with an equable distribution of rain through the seasons, and a moderate and regular inclination of surface—the whole earth, unless rescued by human art from the physical degradation to which it tends, becomes an assemblage of bald mountains, of barren, turfless hills, and of swampy and malarious plains. There are parts of Asia Minor, of Northern Africa, of Greece, and even of Alpine Europe, where the operation of causes set in action by man has brought the face of the earth to a desolation almost as complete as that of the moon; and though, within that brief space of time which we call "the historical period," they are known to have been covered with luxuriant woods, verdant pastures, and fertile meadows, they are now too far deteriorated to be reclaimable by man, nor can they become again fitted for human use, except through great geological changes, or other mysterious influences or agencies of which we have no present knowledge, and over which we have no prospective control. The earth is fast becoming an unfit home for its noblest inhabitant, and another era of equal human crime and human improvidence, and of like duration with that through which traces of that crime and that improvidence extend, would reduce it to such a condition of impoverished productiveness, of shattered surface, of climatic excess, as to threaten the depravation, barbarism, and perhaps even extinction of the species.[5]

Physical Improvement.

True, there is a partial reverse to this picture. On narrow theatres, new forests have been planted; inundations of flowing streams restrained by heavy walls of masonry and other constructions; torrents compelled to aid, by depositing the slime with which they are charged, in filling up lowlands, and raising the level of morasses which their own overflows had created; ground submerged by the encroachments of the ocean, or exposed to be covered by its tides, has been rescued from its dominion by diking; swamps and even lakes have been drained, and their beds brought within the domain of agricultural industry; drifting coast dunes have been checked and made productive by plantation; seas and inland waters have been repeopled with fish, and even the sands of the Sahara have been fertilized by artesian fountains. These achievements are more glorious than the proudest triumphs of war, but, thus far, they give but faint hope that we shall yet make full atonement for our spendthrift waste of the bounties of nature.[6]

Limits of Human Power.

It is, on the one hand, rash and unphilosophical to attempt to set limits to the ultimate power of man over inorganic nature, and it is unprofitable, on the other, to speculate on what may be accomplished by the discovery of now unknown and unimagined natural forces, or even by the invention of new arts and new processes. But since we have seen aerostation, the motive power of elastic vapors, the wonders of modern telegraphy, the destructive explosiveness of gunpowder, of nitro-glycerine, and even of a substance so harmless, unresisting, and inert as cotton, there is little in the way of mechanical achievement which seems hopelessly impossible, and it is hard to restrain the imagination from wandering forward a couple of generations to an epoch when our descendants shall have advanced as far beyond us in physical conquest, as we have marched beyond the trophies erected by our grandfathers. There are, nevertheless, in actual practice, limits to the efficiency of the forces which we are now able to bring into the field, and we must admit that, for the present, the agencies known to man and controlled by him are inadequate to the reducing of great Alpine precipices to such slopes as would enable them to support a vegetable clothing, or to the covering of large extents of denuded rock with earth, and planting upon them a forest growth. Yet among the mysteries which science is hereafter to reveal, there may be still undiscovered methods of accomplishing even grander wonders than these. Mechanical philosophers have suggested the possibility of accumulating and treasuring up for human use some of the greater natural forces, which the action of the elements puts forth with such astonishing energy. Could we gather, and bind, and make subservient to our control, the power which a West Indian hurricane exerts through a small area in one continuous blast, or the momentum expended by the waves, in a tempestuous winter, upon the breakwater at Cherbourg,[7] or the lifting power of the tide, for a month, at the head of the Bay of Fundy, or the pressure of a square mile of sea water at the depth of five thousand fathoms, or a moment of the might of an earthquake or a volcano, our age—which moves no mountains and casts them into the sea by faith alone—might hope to scarp the rugged walls of the Alps and Pyrenees and Mount Taurus, robe them once more in a vegetation as rich as that of their pristine woods, and turn their wasting torrents into refreshing streams.[8]

Could this old world, which man has overthrown, be rebuilded, could human cunning rescue its wasted hillsides and its deserted plains from solitude or mere nomade occupation, from barrenness, from nakedness, and from insalubrity, and restore the ancient fertility and healthfulness of the Etruscan sea coast, the Campagna and the Pontine marshes, of Calabria, of Sicily, of the Peloponnesus and insular and continental Greece, of Asia Minor, of the slopes of Lebanon and Hermon, of Palestine, of the Syrian desert, of Mesopotamia and the delta of the Euphrates, of the Cyrenaica, of Africa proper, Numidia, and Mauritania, the thronging millions of Europe might still find room on the Eastern continent, and the main current of emigration be turned towards the rising instead of the setting sun.

But changes like these must await not only great political and moral revolutions in the governments and peoples by whom those regions are now possessed, but, especially, a command of pecuniary and of mechanical means not at present enjoyed by those nations, and a more advanced and generally diffused knowledge of the processes by which the amelioration of soil and climate is possible than now anywhere exists. Until such circumstances shall conspire to favor the work of geographical regeneration, the countries I have mentioned, with here and there a local exception, will continue to sink into yet deeper desolation, and in the meantime the American continent, Southern Africa, Australia, New Zealand, and the smaller oceanic islands, will be almost the only theatres where man is engaged, on a great scale, in transforming the face of nature.

Importance of Physical Conservation and Restoration.

Comparatively short as is the period through which the colonization of foreign lands by European emigrants extends, great and, it is to be feared, sometimes irreparable injury has already been done in the various processes by which man seeks to subjugate the virgin earth; and many provinces, first trodden by the *homo sapiens Europæ* within the last two centuries, begin to show signs of that melancholy dilapidation which is now driving so many of the peasantry of Europe from their native hearths. It is evidently a matter of great moment, not only to the population of the states where these symptoms are manifesting themselves, but to the general interests of humanity, that this decay should be arrested, and that the future operations of rural husbandry and of forest industry, in districts yet remaining substantially in their native condition, should be so conducted as to prevent the widespread mischiefs which have been elsewhere produced by thoughtless or wanton destruction of the natural safeguards of the soil. This can be done only by the diffusion of knowledge on this subject among the classes that, in earlier days, subdued and tilled ground in which they had no vested rights, but who, in our time, own their woods, their pastures, and their ploughlands as a perpetual possession for them and theirs, and have, therefore, a strong interest in the protection of their domain against deterioration.

Physical Restoration.

Many circumstances conspire to invest with great present interest the questions: how far man can permanently modify and ameliorate those physical conditions of terrestrial surface and climate on which his material welfare depends; how far he can compensate, arrest, or retard the deterioration which many of his agricultural and industrial processes tend to produce; and how far he can restore fertility and salubrity to soils which his follies or his crimes have made barren or pestilential. Among these circumstances, the most prominent, perhaps, is the necessity of providing new homes for a European population which is increasing more rapidly than its means of subsistence, new physical comforts for classes of the people that have now become too much enlightened and have imbibed too much culture to submit to a longer deprivation of a share in the material enjoyments which the privileged ranks have hitherto monopolized.

To supply new hives for the emigrant swarms, there are, first, the vast unoccupied prairies and forests of America, of Australia, and of many other great oceanic islands, the sparsely inhabited and still unexhausted soils of Southern and even Central Africa, and, finally, the impoverished and half-depopulated shores of the Mediterranean, and the interior of Asia Minor and the farther East. To furnish to those who shall remain after emigration shall have conveniently reduced the too dense population of many European states, those means of sensuous and of intellectual well-being which are styled "artificial wants" when demanded by the humble and the poor, but are admitted to be "necessaries" when claimed by the noble and the rich, the soil must be stimulated to its highest powers of production, and man's utmost ingenuity and energy must be tasked to renovate a nature drained, by his improvidence, of fountains which a wise economy would have made plenteous and perennial sources of beauty, health, and wealth.

In those yet virgin lands which the progress of modern discovery in both hemispheres has brought and is still bringing to the knowledge and control of civilized man, not much improvement of great physical conditions is to be looked for. The proportion of forest is indeed to be considerably reduced, superfluous waters to be drawn off, and routes of internal communication to be constructed; but the primitive geographical and climatic features of these countries ought to be, as far as possible, retained.

In reclaiming and reoccupying lands laid waste by human improvidence or malice, and abandoned by man, or occupied only by a nomade or thinly scattered population, the task of the pioneer settler is of a very different character. He is to become a co-worker with nature in the reconstruction of the damaged fabric which the negligence or the wantonness of former lodgers has rendered untenantable. He must aid her in reclothing the mountain slopes with forests and vegetable mould, thereby restoring the fountains which she provided to water them; in checking the devastating fury of torrents, and bringing back the surface drainage to its primitive narrow channels; and in drying deadly morasses by opening the natural sluices which have been choked up, and cutting new canals for drawing off their stagnant waters. He must thus, on the one hand, create new reservoirs, and, on the other, remove mischievous accumulations of moisture, thereby equalizing and regulating the sources of atmospheric humidity and of flowing water, both which are so essential to all vegetable growth, and, of course, to human and lower animal life.

I have remarked that the effects of human action on the forms of the earth's surface could not always be distinguished from those resulting from geological causes, and there is also much uncertainty in respect to the precise influence of the clearing and cultivating of the ground, and of other rural operations, upon climate. It is disputed whether either the mean or the extremes of temperature, the periods of the seasons, or the amount or distribution of precipitation and of evaporation, in any country whose annals are known, have undergone any change during the historical period. It is, indeed, as has been already observed, impossible to doubt that many of the operations of the pioneer settler *tend* to produce great modifications in atmospheric humidity, temperature, and electricity; but we are at present unable to determine how far one set of effects is neutralized by another, or compensated by unknown agencies. This question scientific research is inadequate to solve, for want of the necessary data; but well conducted observation, in regions now first brought under the occupation of man, combined with such historical evidence as still exists, may be expected at no distant period to throw much light on this subject.

Australia and New Zealand are, perhaps, the countries from which we have a right to expect the fullest elucidation of these difficult and disputable problems. Their colonization did not commence until the physical sciences had become matter of almost universal attention, and is, indeed, so recent that the memory of living men embraces the principal epochs of their history; the peculiarities of their fauna, their flora, and their geology are such as to have excited for them the liveliest interest of the votaries of natural science; their mines have given their people the necessary wealth for procuring the means of instrumental observation, and the leisure required for the pursuit of scientific research; and large tracts of virgin forest and natural meadow are rapidly passing under the control of civilized man. Here, then, exist greater facilities and stronger motives for the careful study of the topics in question than have ever been found combined in any other theatre of European colonization.

In North America, the change from the natural to the artificial condition of terrestrial surface began about the period when the most important instruments of meteorological observation were invented. The first settlers in the territory now constituting the United States and the British American provinces had other things to do than to tabulate barometrical and thermometrical readings, but there remain some interesting physical records from the early days of the colonies,[9] and there is still an immense extent of North American soil where the industry and the folly of man have as yet produced little appreciable change. Here, too, with the present increased facilities for scientific observation, the future effects, direct and contingent, of man's labors, can be measured, and such precautions taken in those rural processes which we call improvements, as to mitigate evils, perhaps, in some degree, inseparable from every attempt to control the action of natural laws.

In order to arrive at safe conclusions, we must first obtain a more exact knowledge of the topography, and of the prevalent superficial and climatic condition of countries where the natural surface is as yet more or less unbroken. This can only be accomplished by accurate surveys, and by a great multiplication of the points of meteorological registry,[10] already so numerous; and as, moreover, considerable changes in the proportion of forest and of cultivated land, or of dry and wholly or partially submerged surface, will often take place within brief periods, it is highly desirable that the attention of observers, in whose neighborhood the clearing of the soil, or the drainage of lakes and swamps, or other great works of rural improvement, are going on or meditated, should be especially drawn not only to revolutions in atmospheric temperature and precipitation, but to the more easily ascertained and perhaps more important local changes produced by these operations in the temperature and the hygrometric state of the superficial strata of the earth, and in its spontaneous vegetable and animal products.

The rapid extension of railroads, which now everywhere keep pace with, and sometimes even precede, the occupation of new soil for agricultural purposes, furnishes great facilities for enlarging our knowledge of the topography of the territory they traverse, because their cuttings reveal the composition and general structure of surface, and the inclination and elevation of their lines constitute known hypsometrical sections, which give numerous points of departure for the measurement of higher and lower stations, and of course for determining the relief and depression of surface, the slope of the beds of watercourses, and many other not less important questions.[11]

The geological, hydrographical, and topographical surveys, which almost every general and even local government of the civilized world is carrying on, are making yet more important contributions to our stock of geographical and general physical knowledge, and, within a comparatively short space, there will be an accumulation of well established constant and historical facts, from which we can safely reason upon all the relations of action and reaction between man and external nature.

But we are, even now, breaking up the floor and wainscoting and doors and window frames of our dwelling, for fuel to warm our bodies and to seethe our pottage, and the world cannot afford to wait till the slow and sure progress of exact science has taught it a better economy. Many practical lessons have been learned by the common observation of unschooled men; and the teachings of simple experience, on topics where natural philosophy has scarcely yet spoken, are not to be despised.

In these humble pages, which do not in the least aspire to rank among scientific expositions of the laws of nature, I shall attempt to give the most important practical conclusions suggested by the history of man's efforts to replenish the earth and subdue it; and I shall aim to support those conclusions by such facts and illustrations only as address themselves to the understanding of every intelligent reader, and as are to be found recorded in works capable of profitable perusal, or at least consultation, by persons who have not enjoyed a special scientific training.

NOTES

1. The terrible destructiveness of man is remarkably exemplified in the chase of large mammalia and birds for single products, attended with the entire waste of enormous quantities of flesh, and of other parts of the animal which are capable of valuable uses. The wild cattle of South America are slaughtered by millions for their hides and horns; the buffalo of North America for his skin or his tongue; the elephant, the walrus, and the narwhal for their tusks; the cetacea, and some other marine animals, for their whalebone and oil; the ostrich and other large birds, for their plumage. Within a few years, sheep have been killed in New England, by whole flocks, for their pelts and suet alone, the flesh being thrown away; and it is even said that the bodies of the same quadrupeds have been used in Australia as

fuel for limekilns. What a vast amount of human nutriment, of bone, and of other animal products valuable in the arts, is thus recklessly squandered! In nearly all these cases, the part which constitutes the motive for this wholesale destruction, and is alone saved, is essentially of insignificant value as compared with what is thrown away. The horns and hide of an ox are not economically worth a tenth part as much as the entire carcass. During the present year large quantities of Indian corn have been used as domestic fuel, and even for burning lime, in Iowa and other Western States. Corn at from fifteen to eighteen cents per bushel is found cheaper than wood at from five to seven dollars per cord, or coal at six or seven dollars per ton.—*Rep. Agric. Dept.*, Nov. and Dec., 1872, p. 487.

One of the greatest benefits to be expected from the improvements of civilization is, that increased facilities of communication will render it possible to transport to places of consumption much valuable material that is now wasted because the price at the nearest market will not pay freight. The cattle slaughtered in South America for their hides would feed millions of the starving population of the Old World, if their flesh could be economically preserved and transported across the ocean. This, indeed, is already done, but on a scale which, though absolutely considerable, is relatively insignificant. South America sends to Europe a certain quantity of nutriment in the form of meat extracts, Liebig's and others; and preserved flesh from Australia is beginning to figure in the English market.

We are beginning to learn a better economy in dealing with the inorganic world. The utilization—or, as the Germans more happily call it, the Verwerthung, the *beworthing*—of waste from metallurgical, chemical, and manufacturing establishments, is among the most important results of the application of science to industrial purposes. The incidental products from the laboratories of manufacturing chemists often become more valuable than those for the preparation of which they were erected. The slags from silver refineries, and even from smelting houses of the coarser metals, have not unfrequently yielded to a second operator a better return than the first had derived from dealing with the natural ore; and the saving of lead carried off in the smoke of furnaces has, of itself, given a large profit on the capital invested in the works. According to *Ure's Dictionary of Arts*, see vol. ii., p. 832, an English miner has constructed flues five miles in length for the condensation of the smoke from his lead-works, and makes thereby an annual saving of metal to the value of ten thousand pounds sterling. A few years ago, an officer of an American mint was charged with embezzling gold committed to him for coinage. He insisted, in his defence, that much of the metal was volatilized and lost in refining and melting, and upon scraping the chimneys of the melting furnaces and the roofs of the adjacent houses, gold enough was found in the soot to account for no small part of the deficiency.

The substitution of expensive machinery for manual labor, even in agriculture—not to speak of older and more familiar applications—besides being highly remunerative, has better secured the harvests, and it is computed that the 230,000 threshing machines used in the United States in 1870 obtained five per cent. more grain from the sheaves which passed through them than could have been secured by the use of the flail.

The cotton growing States in America produce annually nearly three million tons of cotton seed. This, until very recently, has been thrown away as a useless incumbrance, but it is now valued at ten or twelve dollars per ton for the cotton fibre which adheres to it, for the oil extracted from it, and for the feed which the refuse furnishes to cattle. The oil—which may be described as neutral—is used very largely for mixing with other oils, many of which bear a large proportion of it without injury to their special properties.

There are still, however, cases of enormous waste in many mineral and mechanical industries. Thus, while in many European countries common salt is a government monopoly, and consequently so dear that the poor do not use as much of it as health requires, in others, as in Transylvania, where it is quarried like stone, the large blocks only are saved, the fragments, to the amount of millions of hundred weights, being thrown away.—BONAR, *Transylvania*, p. 455, 6.

One of the most interesting and important branches of economy at the present day is the recovery of agents such as ammonia and others which had been utilized in chemical manufactures, and re-employing them indefinitely afterwards in repeating the same process.

Among the supplemental exhibitions which will be formed in connection with the Vienna Universal Exhibition is to be one showing what steps have been taken since 1851 (the date of the first London Exhibition) in the utilization of substances previously regarded as waste. On the one hand will be shown the waste products in all the industrial processes included in the forthcoming Exhibition; on the other hand, the useful products which have been obtained from such wastes since 1851. This is intended to serve as an incentive to further researches in the same important direction.

2. It is an interesting and not hitherto sufficiently noticed fact, that the domestication of the organic world, so far as it has yet been achieved, belongs, not indeed to the savage state, but to the earliest dawn of civilization, the conquest of inorganic nature almost as exclusively to the most advanced stages of artificial culture. Civilization has added little to the number of vegetable or animal species grown in our fields or bred in our folds—the cranberry and the wild grape being almost the only plants which the Anglo-American has reclaimed out of our vast native flora and added to his harvests—while, on the contrary, the subjugation of the inorganic forces, and the consequent extension of man's sway over, not the annual products of the earth only, but her substance and her springs of action, is almost entirely the work of highly refined and cultivated ages. The employment of the elasticity of wood and of horn, as a projectile power in the bow, is nearly universal among the, rudest savages. The application of compressed air to the same purpose, in the blowpipe, is more restricted, and the use of the mechanical powers, the inclined plane, the wheel and axle, and even the wedge and lever, seems almost unknown except to civilized man. I have myself seen European peasants to whom one of the simplest applications of this latter power was a revelation.

It is familiarly known to all who have occupied themselves with the psychology and habits of the ruder races, and of persons with imperfectly developed intellects in civilized life, that although these humble tribes and individuals sacrifice, without scruple, the lives of the lower animals to the gratification of their appetites and the supply of their other physical wants, yet they nevertheless seem to cherish with brutes, and even with vegetable life, sympathies which are much more feebly felt by civilized men. The popular traditions of the simpler peoples recognize a certain community of nature between man, brute animals, and even plants; and this serves to explain why the apologue or fable, which ascribes the power of speech and the faculty of reason to birds, quadrupeds, insects, flowers, and trees, is one of the earliest forms of literary composition.

In almost every wild tribe, some particular quadruped or bird, though persecuted as a destroyer of other animals more useful to man, or hunted for food, is regarded with peculiar respect, one might almost say, affection. Some of the North American aboriginal nations celebrate a propitiatory feast to the manes of the intended victim before they commence a bear hunt; and the Norwegian peasantry have not only retained an old proverb which ascribes to the same animal "ti Mœnds Styrke og tolv Mœnds Vid," ten men's strength and twelve men's cunning, but they still pay to him something of the reverence with which ancient superstition invested him. The student of Icelandic literature will find in the saga of Finnbogi hinn rami a curious illustration of this feeling, in an account of a dialogue between a Norwegian bear and an Icelandic champion—dumb show on the part of Bruin, and chivalric words on that of Finnbogi—followed by a duel, in which the latter, who had thrown away his arms and armor in order that the combatants might meet on equal terms, was victorious. See also FRISS, *Lappisk Mthologi*, Christiania, 1871, §37, and the earlier authors there cited. Drummond Hay's very interesting work on Morocco contains many amusing notices of a similar feeling entertained by the Moors towards the redoubtable enemy of their flocks—the lion.

This sympathy helps us to understand how it is that most if not all the domestic animals—if indeed they ever existed in a wild state—were appropriated, reclaimed and trained before men had been gathered into organized and fixed communities, that almost every known esculent plant had acquired substantially its present artificial character, and that the properties of nearly all vegetable drugs and poisons were known at the remotest period to which historical records reach. Did nature bestow upon primitive man some instinct akin to that by which she has been supposed to teach the brute to select the nutritious and to reject the noxious vegetables indiscriminately, mixed in forest and pasture?

This instinct, it must be admitted, is far from infallible, and, as has been hundreds of times remarked by naturalists, it is in many cases not an original faculty but an acquired and transmitted habit. It is a fact familiar to persons engaged in sheep husbandry in New England—and I have seen it confirmed by personal observation—that sheep bred where the common laurel, as it is called, *Kalmia angustifolia*, abounds, almost always avoid browsing upon the leaves of that plant, while those brought from districts where laurel is unknown, and turned into pastures where it grows, very often feed upon it and are poisoned by it. A curious acquired and hereditary instinct, of a different character, may not improperly be noticed here. I refer to that by which horses bred in provinces where quicksands are common avoid their dangers or extricate themselves from them. See BRÉMONTIER, *Mémoire sur de las Dunes, Annales des Ponts et Chaussées*, 1833.: *premier sémestre*, pp. 155–157.

It is commonly said in New England, and I believe with reason, that the crows of this generation are wiser than their ancestors. Scarecrows which were effectual fifty years ago are no longer respected by the plunderers of the cornfield, and new terrors must from time to time be invented for its protection.

Schroeder van der Kolk, in *Het Verschil tusschen den Psychischen Aanleg van het Dier en van den Mensch*, cites many interesting facts respecting instincts lost, or newly developed and become hereditary, in the lower animals, and he quotes Aristotle and Pliny as evidence that the common quadrupeds and fowls of our fields and our poultry yards were much less perfectly domesticated in their times than long, long ages of servitude have now made them.

Among other instances of obliterated instincts, this author states that in Holland, where, for centuries, the young of the cow has been usually taken from the dam at birth and fed by hand, calves, even if left with the mother, make no attempt to suck; while in England, where calves are not weaned until several weeks old, they resort to the udder as naturally as the young of wild quadrupeds.—*Ziel en Ligchaam*, p. 128, *n*.

Perhaps the half-wild character ascribed by P. Læstadius and other Swedish writers to the reindeer of Lapland, may be in some degree due to the comparative shortness of the period during which he has been partially tamed. The domestic swine bred in the woods of Hungary and the buffalo of Southern Italy are so wild and savage as to be very dangerous to all but their keepers. The former have relapsed into their original condition, the latter, perhaps, have never been fully reclaimed from it.

3. The difference between the relations of savage life, and of incipient civilization, to nature, is well seen in that part of the valley of the Mississippi which was once occupied by the mound builders and afterwards by the far less developed Indian tribes. When the tillers of the fields, which must have been cultivated to sustain the large population that once inhabited those regions, perished, or were driven out, the soil fell back to the normal forest state, and the savages who succeeded the more advanced race interfered very little, if at all, with the ordinary course of spontaneous nature.

4. Whatever may be thought of the modification of organic species by natural selection, there is certainly no evidence that animals have exerted upon any form of life an influence analogous to that of domestication upon plants, quadrupeds, and birds reared artificially by man; and this is as true of unforeseen as of purposely effected improvements accomplished by voluntary selection of breeding animals.

   It is true that nature employs birds and quadrupeds for the dissemination of vegetable and even of animal species. But when the bird drops the seed of a fruit it has swallowed, and when the sheep transports in its fleece the seed-vessel of a burdock from the plain to the mountain, its action is purely mechanical and unconscious, and does not differ from that of the wind in producing the same effect.

5. "And it may be remarked that, as the world has passed through these several stages of strife to produce a Christendom, so by relaxing in the enterprises it has learnt, does it tend downwards, through inverted steps, to wildness and the waste again. Let a people give up their contest with moral evil; disregard the injustice, the ignorance, the greediness, that may prevail among them, and part more and more with the Christian element of their civilization; and in declining this battle with sin, they will inevitably get embroiled with men. Threats of war and revolution punish their unfaithfulness; and if then, instead of retracing their steps, they yield again, and are driven before the storm, the very arts they had created, the structures they had raised, the usages they had established, are swept away; 'in that very day their thoughts perish.' The portion they had reclaimed from the young earth's ruggedness is lost; and failing to stand fast against man, they finally get embroiled with nature, and are thrust down beneath her ever-living hand."—MARTINEAU'S *Sermon, "The Good Soldier of Jesus Christ."*

6. The wonderful success which has attended the measures for subduing torrents and preventing inundations employed in Southern France since 1865, and described in Chapter III., *post*, ought to be here noticed as a splendid and most encouraging example of well-directed effort in the way of physical restoration.

7. In heavy storms, the force of the waves as they strike against a sea-wall is from one and a half to two tons to the square foot, and Stevenson, in one instance at Skerryvore and In another at the Bell Rock lighthouse, found this force equal to nearly three tons per foot.

The seaward front of the breakwater at Cherbourg exposes a surface of about 2,500,000 square feet. In rough weather the waves beat against this whole face, though at the depth of twenty-two yards, which is the height of the breakwater, they exert a very much less violent motive force than at and near the surface of the sea, because this force diminishes in geometrical, as the distance below the surface increases in arithmetical, proportion. The shock of the waves is received several thousand times in the course of twenty-four hours, and hence the sum of impulse which the breakwater resists in one stormy day amounts to many thousands of millions of tons. The breakwater is entirely an artificial construction. If then man could accumulate and control the forces which he is able effectually to resist, he might be said to be, physically speaking, omnipotent.

8. Some well-known experiments show that it is quite possible to accumulate the solar heat by a simple apparatus, and thus to obtain a temperature which might be economically important even in the climate of Switzerland. Saussure, by receiving the sun's rays in a nest of bores blackened within and covered with glass, raised a thermometer enclosed in the Inner box, to the boiling point; and under the more powerful sun of the Cape of Good Hope, Sir John Herschel cooked the materials for a family dinner by a similar process, using, however, but a single box, surrounded with dry sand and covered with two glasses. Why should not so easy a method of economizing fuel be resorted to in Italy, in Spain, and even in more northerly climates?

   The unfortunate John Davidson records in his journal that he saved fuel in Morocco by exposing his teakettle to the sun on the roof of his house, where the water rose to the temperature of one hundred and forty degrees, and, of course, needed little fire to bring it to boil. But this was the direct and simple, not the concentrated or accumulated heat of the sun.

   On the utilizing of the solar heat, simply as heat, see the work of MOUCHOT, *La Chaleur solaire et ses applications industrielles.* Paris, 1869.

   The reciprocal convertibility of the natural forces has suggested the possibility of advantageously converting the heat of the sun into mechanical power. Ericsson calculates that in all latitudes between the equator and 45°, a hundred square feet of surface exposed to the solar rays develop continuously, for nine hours a day on an average, eight and one fifth horse-power.

   I do not know that any attempts have been made to accumulate and store up, for use at pleasure, force derived from this powerful source.

9. The Travels of Dr. Dwight, president of Yale College, which embody the results of his personal observations, and of his inquiries among the early settlers, in his vacation excursions in the Northern States of the American Union, though presenting few instrumental measurements or tabulated results, are of value for the powers of observation they exhibit, and for the sound common sense with which many natural phenomena, such for instance as the formation of the river meadows, called "intervales," in New England, are explained. They present a true and interesting picture of physical conditions, many of which have long ceased to exist in the theatre of his researches, and of which few other records are extant.

10. The general law of temperature is that it decreases as we ascend. But, in hilly regions, the law is reversed in cold, still weather, the cold air descending, by reason of its greater gravity, into the valleys. If there be wind enough, however, to produce a disturbance and intermixture of higher and lower atmospheric strata, this exception to the general law does not take place. These facts have long been familiar to the common people of Switzerland and of New England, but their

importance has not been sufficiently taken into account in the discussion of meteorological observations. The descent of the cold air and the rise of the warm affect the relative temperatures of hills and valleys to a much greater extent than has been usually supposed. A gentleman well known to me kept a thermometrical record for nearly half a century, in a New England country town, at an elevation of at least 1,500 feet above the sea. During these years his thermometer never fell lower than 26°— Fahrenheit, while at the shire town of the county, situated in a basin one thousand feet lower, and only ten miles distant, as well as at other points in similar positions, the mercury froze several times in the same period.

11. Railroad surveys must be received with great caution where any motive exists for cooking them. Capitalists are shy of investments in roads with steep grades, and of course it is important to make a fair show of facilities in obtaining funds for new routes. Joint-stock companies have no souls; their managers, in general, no consciences. Cases can be cited where engineers and directors of railroads, with long grades above one hundred feet to the mile, have regularly sworn in their annual reports, for years in succession, that there were no grades upon their routes exceeding half that elevation. In fact, every person conversant with the history of these enterprises knows that in their public statements falsehood is the rule, truth the exception.

What I am about to remark is not exactly relevant to my subject; but it is hard to "get the floor" in the world's great debating society, and when a speaker who has anything to say once finds access to the public ear, he must make the most of his opportunity, without inquiring too nicely whether his observations are "in order." I shall harm no honest man by endeavoring, as I have often done elsewhere, to excite the attention of thinking and conscientious men to the dangers which threaten the great moral and even political interests of Christendom, from the unscrupulousness of the private associations that now control the monetary affairs, and regulate the transit of persons and property, in almost every civilized country. More than one American State is literally governed by unprincipled corporations, which not only defy the legislative power, but have, too often, corrupted even the administration of justice. The tremendous power of these associations is due not merely to pecuniary corruption, but partly to an old legal superstition—fostered by the decision of the Supreme Court of the United States in the famous Dartmouth College case—in regard to the sacredness of corporate prerogatives. There is no good reason why private *rights* derived from God and the very constitution of society should be less respected than privileges granted by legislatures. It should never be forgotten that no *privilege* can be a *right*, and legislative bodies ought never to make a grant to a corporation, without express reservation of what many sound jurists now hold to be involved in the very nature of such grants, the power of revocation. Similar evils have become almost equally rife in England, and on the Continent; and I believe the decay of commercial morality, and of the sense of all higher obligations than those of a pecuniary nature, on both sides of the Atlantic, is to be ascribed more to the influence of joint-stock banks and manufacturing and railway companies, to the workings, in short, of what is called the principle of "associate action," than to any other one cause of demoralization.

The apophthegm, "the world is governed too much," though unhappily too truly spoken of many countries—and perhaps, in some aspects, true of all—has done much mischief whenever it has been too unconditionally accepted as a political axiom. The popular apprehension of being over-governed, and, I am afraid, more emphatically the fear of being over-taxed, has had much to do with the general abandonment of certain governmental duties by the ruling powers of most modern states. It is theoretically the duty of government to provide all those public facilities of intercommunication and commerce, which are essential to the prosperity of civilized commonwealths, but which individual means are inadequate to furnish, and for the due administration of which individual guaranties are insufficient. Hence public roads, canals, railroads postal communications, the circulating medium of exchange whether metallic or representative, armies, navies, being all matters in which the nation at large has a vastly deeper interest than any private association can have, ought legitimately to be constructed and provided only by that which is the visible personification and embodiment of the nation, namely, its legislative head. No doubt the organization and management of these institutions by government are liable, as are all things human, to great abuses. The multiplication of public placeholders, which they imply, is a serious evil. But the corruption thus engendered, foul as it is, does not strike so deep as the rottenness of private corporations; and official rank, position, and duty have, in practice, proved better securities for fidelity and pecuniary integrity in the conduct of the interests in question, than the suretyships of private corporate agents, whose bondsmen so often fail or abscond before their principal is detected.

Many theoretical statesmen have thought that voluntary associations for strictly pecuniary and industrial purposes, and for the construction and control of public works, might furnish, in democratic countries, a compensation for the small and doubtful advantages, and at the same time secure an exemption from the great and certain evils, of aristocratic institutions. The example of the American States shows that private corporations—whose rule of action is the interest of the association, not the conscience of the individual—though composed of ultra-democratic elements, may become most dangerous enemies to rational liberty, to the moral interests of the commonwealth, to the purity of legislation and of judicial action, and to the sacredness of private rights.

12

William Cullen Bryant (1794–1878)

While still in his teens, poems such as "Thanatopsis" and "To a Waterfowl" gained Bryant a reputation as one of America's most promising poets. Much of Bryant's poetry reflects his love of the outdoors and his belief that nature is the visible manifestation of God. In 1825 he became coeditor of the New York Evening Post, *and a few years later he became the editor-in-chief, a position he would hold for nearly fifty years. From the mid to late 1800s, Bryant's love of nature was manifested in numerous articles and editorials supporting the creation of parks and forest preserves throughout the United States. As can be seen from this article, Bryant was also one of the first to recognize the importance of Marsh's* Man and Nature *and to incorporate Marsh's rationales for forest preservation into his own arguments in favor of creating forest preserves.*

"The Utility of Trees" (1865)

While Congress is occupied with the disposition of the public lands, it has been suggested, by persons who can think of something else besides railroads, that it will be an act of provident wisdom to reserve considerable tracts of forest in different parts of the country, as the public domain, with a view of preventing the destruction of trees which is so rapidly proceeding, and which may yet lead to inconvenient consequences. One of the objections to any such measure as this is the difficulty of protecting the forests from depredation. What is the public domain is regarded in most parts of our country as, in a certain sense, common property, from which any man may take what he has occasion for. In order to prevent the trees from being felled and the public property wasted, a body of foresters to watch it and keep out trespassers must be retained in the pay of the Government.

We do not propose here to discuss the merits of any such plan, but we would refer our readers to the extracts we have made in another part of this

New York *Evening Post*, June 20, 1865.

paper from Dr. Piper's work on the "Trees of America," as opening a very important subject for their consideration. We are fully persuaded, for our part, that scarce anything is more prejudicial to the fertility of a country, or has a worse effect on its climate, than the thoughtless practice of denuding it of trees. The effect is to open the region to the winds which parch the surface and dwarf the vegetation, to cause the springs to dry up, and turn the water-courses into torrents in the rainy season and dusty channels in summer and autumn. If the public could but be thoroughly convinced of this truth, it might, perhaps, be unnecessary for the Government to give itself any concern about the matter in the disposition of the public lands. We should see the highways skirted by double rows of trees, long lines of plantation following the courses of the railroads, belts of forest-trees planted to break the force of the winds and shelter the tender crops and the orchards which bear fruit.

Travellers see in the Orkney and Shetland Isles a remarkable example of the unfavorable effect of the winds on vegetation, and the kindly influence of shelter. There, within the high garden-walls, shrubs and trees grow and flourish till their summits reach the top of the wall, and beyond that they do not rise. Strong winds sweep almost constantly over the surface of the islands and prevent the growth of any vegetable production except those which flower and perfect their fruit near the surface. Within the enclosures of these stone fences, where the wind is excluded, or in hollows which form a natural screen, the vegetation is often juicy and luxuriant. Of the effect of the destruction of forests in drying up a country the Old World affords many remarkable instances. The poets and historians of antiquity speak of mighty forests which are now shadowless wastes, and of streams that flow no longer. Addison thus laments the disappearance of rivers and brooks celebrated in ancient times:

> "Sometimes, misguided by the tuneful throng,
> I look for streams immortalized in song
> That lost in silence and oblivion lie;
> Dumb are their fountains, and their channels dry,
> Yet run forever, by the Muses' skill,
> And, in the smooth description, murmur still."

The streams of Attica, now a bare and treeless region, are for the most part but mere water-courses, in which the rains pass off to the sea, and disappoint those who form their ideas of them from the ancient writers.

The rivers of Spain are for the most part only channels for the winter rains. The Guadalquivir, which some poet calls "a mighty river," enters the sea at Malaga without water enough to cover the loose black stones that pave its bed. The Holy Land now often misses "the latter rain," or receives it but sparingly; and the brook Kedron is a long, dry ravine, passing off to the eastward from Jerusalem to descend between perpendicular walls beside the monastery of Mar Saba to the valley of the Jordan and the Dead Sea. Mr. Marsh, in his very instructive book entitled "Man and Nature," has collected a vast number of instances showing how, in the Old World, the destruction of the forests has been followed by a general aridity of the country which they formerly overshadowed.

Whether there are any examples of frequent rains restored to a country by planting groves and orchards we can not say; but we remember, when travelling at the West thirty-three years since, to have met with a gentleman from Kentucky who spoke of an instance within his knowledge in which a perennial stream had made its appearance where at the early settlement of the region there was none. Kentucky, when its first colonists planted themselves within its limits, was a region in which extensive prairies, burnt over every year by the Indians, predominated.

More than forty years since a poet of our country, referring to the effect of stripping the soil of its trees, put these lines into the mouth of one of the aboriginal inhabitants:

> "Before these fields were shorn and tilled,
> Full to the brim our rivers flowed;
> The melody of waters filled
> The fresh and boundless wood;
> And torrents dashed and rivulets played,
> And fountains spurted in the shade.
> "Those grateful sounds are heard no more;
> The springs are silent in the sun;
> The rivers, by the blackened shore,
> With lessening currents run.
> The realm our tribes are crushed to get
> May be a barren desert yet."

In all woodlands nature has made provision for retaining the moisture of rains long in the ground. The earth under the trees is covered with a thick carpeting of fallen leaves, which absorb the power and prevent the water from passing immediately into the streams and hurrying to the sea. Part of the moisture thus confined under the fallen leaves, and shielded from evaporation by sun and wind, finds its way slowly into the veins of the earth, rises in springs, and runs off in rivulets; part is gradually drawn up by the rootlets of the trees and given off to the air from the leaves, to form the vapors, which are afterward condensed into showers. We fear that this statement of the process is somewhat commonplace, but we make it here because, though obvious enough, it is too little in the minds of those whose interest it concerns to notice it. Thus it is that forests protect a country against drought, and keep its streams constantly flowing and its wells constantly full. Cut down the trees, and the moisture of the showers passes rapidly off from the surface and hastens to lakes and to ocean.

13

John Muir (1838–1914)

When Muir was eleven years old, his family moved from Dunbar, Scotland to the Wisconsin prairie, where the family established a farm on the frontier. As Muir later recounted in The Story of My Boyhood and Youth *(1912), his father was a stern Christian fundamentalist who used the bonfires in which they burned the brush cleared from the land as a religious lesson, "comparing their heat with that of hell, and the branches with bad boys." Although Muir loved life on the frontier, he chafed under the austere authoritarianism of his father, and left home in 1860 to study at the University of Wisconsin where he studied geology and botany. After leaving the university in 1863, Muir went to Canada, a decision that was partly motivated by his desire to botanize there and partly by a desire to avoid the Civil War era draft. After recovering from a sawmill accident that nearly cost him his eyesight, Muir embarked on a walking tour from Indianapolis to Florida in September 1867. The journal notes from that trip were published after Muir's death as* A Thousand-Mile Walk to the Gulf *(1916). The following year, Muir sailed to San Francisco, and made his first visit to the Yosemite Valley. In 1869 Muir took a job as a sheepherder in the Sierras, an experience he describes in* My First Summer in the Sierra *(1911).*

The "glorious conversion" that Muir experienced in the Sierras became the impetus behind his thirty-year fight for wilderness preservation and the establishment of national parks and forest preserves. After a falling out with Gifford Pinchot over the issue of resource development in federally protected forest preserves, the conservation movement split into two factions, with Pinchot (and his political mentor Theodore Roosevelt) identified with the "wise use" faction and Muir and his allies comprising the wilderness preservationists. In 1892, Muir was instrumental in the founding of the Sierra Club, serving as its first president. In essay after essay, Muir extolled the beauty of the nation's wilderness areas, arguing vigorously for their protection. Despite his many successes, including the creation of Yosemite, Sequoia, and General Grant National Parks, much of Muir's finest polemical writing came in the ultimately doomed effort to prevent the damming of the Hetch Hetchy Valley. In "Hetch Hetchy Valley" Muir methodically rebuts all of the arguments in favor of damming the valley, then concludes the essay with a condemnation of the dam's proponents that likens them to the money-changers that Christ threw out of the temple. On December 24, 1914, less than a year after President Woodrow Wilson signed the bill that led to the damming of Hetch Hetchy, Muir died from pneumonia.

"God's First Temples: How Shall We Preserve Our Forests?" (1876)

EDS. RECORD-UNION: The forests of coniferous trees growing on our mountain ranges are by far the most destructible of the natural resources of California. Our gold, and silver, and cinnabar are stored in the rocks, locked up in the safest of all banks, so that notwithstanding the world has been making a run upon them for the last twenty-five years, they still pay out steadily, and will probably continue to do so centuries hence, like rivers pouring from perennial mountain fountains. The riches of our magnificent soil-beds are also comparatively safe, because even the most barbarous methods of wildcat farming cannot effect complete destruction, and however great the impoverishment produced, full restoration of fertility is always possible to the enlightened farmer. But our forest belts are being burned and cut down and wasted like a field of unprotected grain, and once destroyed can never be wholly restored even by centuries of persistent and painstaking cultivation.

The practical importance of the preservation of our forests is augmented by their relations to climate, soil and streams. Strip off the woods with their underbrush from the mountain flanks, and the whole State, the lowlands as well as the highlands, would gradually change into a desert. During rainfalls, and when the winter snow was melting, every stream would become a destructive torrent, overflowing its banks, stripping off and carrying away the fertile soils, filling up the lower river channels, and overspreading the lowland fields with detritus to a vastly more destructive degree than all the washings from hydraulic mines concerning which we now hear so much. Dripping forests give rise to moist sheets and currents of air, and the sod of grasses and underbrush thus fostered, together with the roots of trees themselves, absorb and hold back rains and melting snow, yet allowing them to ooze and percolate and flow gently in useful fertilizing streams. Indeed every pine needle and rootlet, as well as fallen trunks and large clasping roots, may be regarded as dams, hoarding the bounty of storm clouds, and dispensing it as blessings all through the summer, instead of allowing it to gather and rush headlong in short-lived devastating floods. Streams taking their rise in deep woods flow unfailingly as those derived from the eternal ice and snow of the Alps. So constant indeed and apparent is the relationship between forests and never-failing springs, that effect is frequently mistaken for cause, it being often

Sacramento Record-Union, February 5, 1876.

asserted that fine forests will grow only along streamsides where their roots are well watered, when in fact the forests themselves produce many of the streams flowing through them.

The main forest belt of the Sierra is restricted to the western flank, and extends unbrokenly from one end of the range to the other at an elevation of from three to eight thousand feet above sea level. The great master-existence of these noble woods is sequoia gigantea, or big tree. Only two species of sequoia are known to exist in the world. Both belong to California, one being found only in the Sierra, the other (sequoia sempervirens) in the Coast Ranges, although no less than five distinct fossil species have been discovered in the tertiary and cretaceous rocks of Greenland. I would like to call attention to this noble tree, with special reference to its preservation. The species extends from the well known Calaveras groves on the north, to the head of Deer creek on the south, near the big bend of the Kern river, a distance of about two hundred miles, at an elevation above sea level of from about five to eight thousand feet. From the Calaveras to the South Fork of King's river it occurs only in small isolated groves, and so sparsely and irregularly distributed that two gaps occur nearly forty miles in width, the one between the Calaveras and Tuolumne groves, the other between those of the Fresno and King's rivers. From King's river the belt extends across the broad, rugged basins of the Kaweah and Tule rivers to its southern boundary on Deer creek, interrupted only by deep, rocky canyons, the width of this portion of the belt being from three to ten miles.

In the northern groves few young trees or saplings are found ready to take the places of the failing old ones, and because these ancient, childless sequoias are the only ones known to botanists, the species has been generally regarded as doomed to speedy extinction, as being nothing more than an expiring remnant of an ancient flora, and that therefore there is no use trying to save it or to prolong its few dying days. This, however, is in the main a mistaken notion, for the Sierra as it now exists never had an ancient flora. All the species now growing on the range have been planted since the close of the glacial period, and the Big Tree has never formed a greater part of these post-glacial forests than it does today, however widely it may have been distributed throughout pre-glacial forests.

In tracing the belt southward, all the phenomena hearing upon its history goes to show that the dominion of Sequoia Gigantea, as King of California trees, is not yet passing away. No tree in the woods seems more firmly established, or more safely settled in accordance with climate and soil. They fill the woods and form the principal tree, growing heartily on solid ledges, along water courses, in the deep, moist soil of meadows, and upon avalanche and glacial debris, with a multitude of thrifty seedlings and saplings crowding around the aged, ready to take their places and rule the woods.

Nevertheless Nature in her grandly deliberate way keeps up a rotation of forest crops. Species develop and die like individuals, animal as well as plant. Man himself will as surely become extinct as sequoia or mastodon, and be at length known only as a fossil. Changes of this kind are, however, exceedingly slow in their movements, and, as far as the lives of individuals are concerned, such changes have no appreciable effect. Sequoia seems scarcely further past prime as a species than its companion firs (*Picea amabilis* and *P. grandis*), and judging from its present condition and its ancient history, as far as I have been able to decipher it, our sequoia will live and flourish gloriously until A.D. 15,000 at least—probably for longer that is, if it be allowed to remain in the hands of Nature.

But waste and pure destruction are already taking place at a terrible rate, and unless protective measures be speedily invented and enforced, in a few years this noblest tree-species in the world will present only a few hacked and scarred remnants. The great enemies of forests are fire and the ax. The destructive effects of these, as compared with those caused by the operations of nature, are instantaneous. Floods undermine and kill many a tree, storm winds bend and break, landslips and avalanches overwhelm whole groves, lightning shatters and burns, but the combined effects of all these amount only to a wholesome beauty-producing culture. Last summer I found some five saw mills located in or near the lower edge of the Sequoia belt, all of which saw more or less of the big tree into lumber. One of these (Hyde's), situated on the north fork of the Kaweah, cut no less than 2,000,000 feet of Sequoia lumber last season. Most of the Fresno big trees are doomed to feed the mills recently erected near them, and a company has been formed by Chas. Converse to cut the noble forest on the south fork of King's river. In these milling operations waste far exceeds use. After the choice young manageable trees have been felled, the woods are cleared of limbs and refuse by burning, and in these clearing fires, made with reference to further operations, all the young seedlings and saplings are destroyed, together with many valuable fallen trees and old trees, too large to be cut, thus effectually cutting off all hopes of a renewal of the forest.

These ravages, however, of mill-fires and mill-axes are small as compared with those of the "sheep-men's" fires. Incredible numbers of sheep are driven to the mountain pastures every summer, and in order to make easy paths and to improve the pastures, running fires are set everywhere to burn off the old logs and underbrush. These fires are far more universal and destructive than would be guessed. They sweep through nearly the entire forest belt of the range from one extremity to the other, and in the dry weather, before the coming on of winter storms, are very destructive to all kinds of young trees, and especially to sequoia, whose loose, fibrous bark catches and burns at once. Excepting the Calaveras, I, last summer, examined every sequoia grove in the range, together with the main belt extending across the basins of Kaweah and Tule, and found everywhere the most deplorable waste from this cause. Indians burn off underbrush to facilitate deer-hunting. Campers of all kinds often permit fires to run, so also do mill-men, but the fires of "sheep-men" probably form more than 90 per cent. of all destructive fires that sweep the woods.

Fire, then, is the arch destroyer of our forests, and sequoia forests suffer most of all. The young trees are most easily fire killed; the old are most easily burned, and the prostrate trunks, which *never rot* and would remain valuable until our tenth centennial, are reduced to ashes.

In European countries, especially in France, Germany, Italy and Austria, the economies of forestry have been carefully studied under the auspices of Government, with the most beneficial results. Whether our loose-jointed Government is really able or willing to do anything in the matter remains to be seen. If our law makers were to discover and enforce any method tending to lessen even in a small degree the destruction going on, they would thus cover a multitude of legislative sins in the eyes of every tree lover. I am satisfied, however, that the question can be intelligently discussed only after a careful survey of our forests has been made, together with studies of the forces now acting upon them.

A law was constructed some years ago making the cutting down of sequoias over sixteen feet in diameter illegal. A more absurd and shortsighted piece of legislation could not be conceived. All the young trees might be cut and burned, and all the old ones might be burned but not cut.

"Hetch Hetchy Valley" from *The Yosemite* (1913)

Yosemite is so wonderful that we are apt to regard it as an exceptional creation, the only valley of its kind in the world; but Nature is not so poor as to have only one of anything. Several other yosemites have been discovered in the Sierra that occupy the same relative positions on the Range and were formed by the same forces in the same kind of granite. One of these, the Hetch Hetchy Valley, is in the Yosemite National Park about twenty miles from Yosemite and is easily accessible to all sorts of travellers by a road and trail that leaves the Big Oak Flat road at Bronson Meadows a few miles below Crane Flat, and to mountaineers by way of Yosemite Creek basin and the head of the middle fork of the Tuolumne.

It is said to have been discovered by Joseph Screech, a hunter, in 1850, a year before the discovery of the great Yosemite. After my first visit to it in the autumn of 1871, I have always called it the "Tuolumne Yosemite," for it is a wonderfully exact counterpart of the Merced Yosemite, not only in its sublime rocks and waterfalls but in the gardens, groves and meadows of its flowery park-like floor. The floor of Yosemite is about 4,000 ft. above the sea; the Hetch Hetchy floor about 3,700 ft. And as the Merced River flows through Yosemite, so does the Tuolumne through Hetch Hetchy. The walls of both are of grey granite, rise abruptly from the floor, are sculptured in the same style and in both every rock is a glacier monument.

Standing boldly out from the south wall is a strikingly picturesque rock called by the Indians, Kolana, the outermost of a group 2,300 ft. high, corresponding with the Cathedral Rocks of Yosemite both in relative position and form. On the opposite side of the Valley, facing Kolana, there is a counterpart of the El Capitan that rises sheer and plain to a height of 1,800 ft., and over its massive brow flows a stream which makes the most graceful fall I have ever seen. From the edge of the cliff to the top of an earthquake talus it is perfectly free in the air for a thousand feet before it is broken into cascades among talus boulders. It is in all its glory in June, when the snow is melting fast, but fades and vanishes toward the end of summer. The only fall I know with which it may fairly be compared is the Yosemite Bridal Veil; but it excels even that favourite fall both in height and airy-fairy beauty and behaviour. Lowlanders are apt to suppose that mountain streams in their wild career over cliffs lose control of themselves and tumble in a noisy chaos of mist and spray. On the contrary, on no part of their travels are they more harmonious and self-controlled. Imagine yourself in Hetch Hetchy on a sunny day in June, standing waist-deep in grass and flowers (as I have often stood), while the great pines sway dreamily with scarcely perceptible motion. Looking northward across the Valley you see a plain, grey granite cliff rising abruptly out of the gardens and groves to a height of 1,800 ft., and in front of it Tueeulala's silvery scarf burning with irised sun-fire. In the first white outburst at the head there is abundance of visible energy, but it is speedily hushed and concealed in divine repose and its tranquil progress to the base of the cliff is like that of a downy feather in a still room. Now observe the fineness and marvellous distinctness of the various sun-illumined fabrics into which the water is woven; they sift and float from form to form down the face of that grand grey rock in so leisurely and unconfused a manner that you can examine their texture, and patterns and tones of colour as you would a piece of embroidery held in the hand. Toward the top of the fall you see groups of booming, comet-like masses, their solid, white heads separate, their tails like combed silk interlacing among delicate grey and purple shadows, ever forming and dissolving, worn out by friction in their rush through the air. Most of these vanish a few hundred feet below the summit, changing to varied forms of cloud-like drapery. Near the bottom the width of the fall has increased from about twenty-five feet to a hundred feet. Here it is composed of yet finer tissues, and is still without a trace of disorder—air, water and sunlight woven into stuff that spirits might wear.

So fine a fall might well seem sufficient to glorify any valley; but here, as in Yosemite, Nature seems in nowise moderate, for a short distance to the eastward of Tueeulala booms and thunders the great Hetch Hetchy Fall, Wapama, so near that you have both of them in full view from the same standpoint. It is the counterpart of the Yosemite Fall, but has a much greater volume of water, is about 1,700 ft. in height, and appears to be nearly vertical, though considerably inclined, and is dashed into huge outbounding bosses of foam on projecting shelves and knobs. No two falls could be more unlike—Tueeulala out in the open sunshine descending like thistledown; Wapama in a jagged, shadowy gorge roaring and thundering, pounding its way like an earthquake avalanche.

Besides this glorious pair there is a broad, massive fall on the main river a short distance above the head of the Valley. Its position is something like that of the Vernal in Yosemite, and its roar as it plunges into a surging trout-pool may be heard a long way, though it is only about twenty feet high. On Ranchena Creek, a large stream, corresponding in position with the Yosemite Tenaya Creek, there is a chain of cascades joined here and there with swift flashing plumes like the one between the Vernal and Nevada Falls, making magnificent shows as they go their glacier-sculptured way, sliding, leaping, hurrahing, covered with crisp clashing spray made glorious with sifting sunshine. And besides all these a few small streams come over the walls at wide intervals, leaping from ledge to ledge with birdlike song and watering many a hiden cliff-garden and fernery, but they are too unshowy to be noticed in so grand a place.

The correspondence between the Hetch Hetchy walls in their trends, sculpture, physical structure, and general arrangement of the main rock-masses and those of the Yosemite Valley has excited the wondering admiration of every observer. We have seen that the El Capitan and Cathedral rocks occupy the same relative positions in both valleys; so also do their Yosemite points and North Domes. Again, that part of the Yosemite north wall immediately to the east of the Yosemite Fall has two horizontal benches, about 500 and 1,500 ft. above the floor, timbered with golden-cup oak. Two benches similarly situated and timbered occur on the same relative portion of the Hetch Hetchy north wall, to the east of Wapama Fall, and on no other. The Yosemite is bounded at the head by the great Half Dome. Hetch Hetchy is bounded in the same way, though its head rock is incomparably less wonderful and sublime in form.

The floor of the Valley is about three and a half miles long, and from a fourth to half a mile wide. The lower portion is mostly a level meadow about a mile long, with the trees restricted to the sides and the river banks, and partially separated from the main, upper, forested portion by a low bar of glacier-polished granite across which the river breaks in rapids.

The principal trees are the yellow and sugar pines, digger pine, incense cedar, Douglas spruce, silver fir, the California and golden-cup oaks, balsam cottonwood, Nuttall's flowering dogwood, alder, maple, laurel, tumion, etc. The most abundant and influential are the great yellow or silver pines like those of Yosemite, the tallest over two hundred feet in height, and the oaks assembled in magnificent groves with massive rugged trunks four to six feet in diameter, and broad, shady, wide-spreading heads. The shrubs forming conspicuous flowery clumps and tangles are manzanita, azalea, spirea, brier-rose, several species of ceanothus, calycanthus, philadelphus, wild cherry, etc.; with abundance of showy and fragrant herbaceous plants growing about them or out in the open in beds by themselves—lilies, Mariposa tulips, brodiaeas, orchids, iris, spraguea, draperia, collomia, collinsia, castilleja, nemophila, larkspur, columbine, goldenrods, sunflowers, mints of many species, honeysuckle, etc. Many fine ferns dwell here also, especially the beautiful and interesting rockferns—pellaea, and cheilanthes of several species—fringing and rosetting dry rock-piles and ledges; woodwardia and asplenium on damp spots with fronds six or seven feet high; the delicate maidenhair in mossy nooks by the falls, and the sturdy, broad-shouldered pteris covering nearly all the dry ground beneath the oaks and pines.

It appears, therefore that Hetch Hetchy Valley, far from being a plain, common, rock-bound meadow, as many who have not seen it seem to suppose, is a grand landscape garden, one of Nature's rarest and most precious mountain temples. As in Yosemite, the sublime rocks of its walls seem to glow with life, whether leaning back in repose or standing erect in thoughtful attitudes, giving welcome to storms and calms alike, their brows in the sky, their feet set in the groves and gay flowery meadows, while birds, bees, and butterflies help the river and waterfalls to stir all the air into music—things frail and fleeting and types of permanence meeting here and blending, just as they do in Yosemite, to draw her lovers into close and confiding communion with her.

Sad to say, this most precious and sublime feature of the Yosemite National Park, one of the greatest of all our natural resources for the uplifting joy and peace and health of the people, is in danger of being dammed and made into a reservoir to help supply San Francisco with water and light, thus flooding it from wall to wall and burying its gardens and groves one or two hundred feet deep. This grossly destructive commercial scheme has long been planned and urged (though water as pure and abundant can be got from sources outside of the people's park, in a dozen different places), because of the comparative cheapness of the dam and of the territory which it is sought to divert from the great uses to which it was dedicated in the Act of 1890 establishing the Yosemite National Park.

The making of gardens and parks goes on with civilization all over the world, and they increase both in size and number as their value is recognized. Everybody needs beauty as well as bread, places to play in and pray in, where Nature may heal and cheer and give strength to body and soul alike. This natural beauty-hunger is made manifest in the little window-sill gardens of the poor, though perhaps only a geranium slip in a broken cup, as well as in the carefully tended rose and lily gardens of the rich, the thousands of spacious city parks and botanical gardens, and in our magnificent National parks—the Yellowstone, Yosemite, Sequoia, etc.—Nature's sublime wonderlands, the admiration and joy of the world. Nevertheless, like anything else worth while, from the very beginning, however well guarded, they have always been subject to attack by despoiling gainseekers and mischief-makers of every degree from Satan to Senators, eagerly trying to make everything immediately and selfishly commercial with schemes disguised in smug-smiling philanthropy, industriously, shampiously crying, "Conservation, conservation, panutilization," that man and beast may be fed and the dear Nation made great. Thus long ago a few enterprising merchants utilized the Jerusalem temple as a place of business instead of a place of prayer, changing money, buying and selling cattle and sheep and doves; and earlier still, the first forest reservation, including only one tree, was likewise despoiled. Ever since the establishment of the Yosemite National Park, strife has been going on around its borders and I suppose this will go on as part of the universal battle between right and wrong, however much its boundaries may be shorn, or its wild beauty destroyed.

The first application to the Government by the San Francisco Supervisors for the commercial use of Lake Eleanor and the Hetch Hetchy Valley was made in 1903, and on 22 December of that year it was denied by the Secretary of the Interior, Mr Hitchcock, who truthfully said:

> Presumably the Yosemite National Park was created such by law because of the natural objects of varying degrees of scenic importance located within its boundaries, inclusive alike of its beautiful small lakes, like Eleanor, and its majestic wonders, like Hetch Hetchy and Yosemite Valley. It is the aggregation of such natural scenic features that makes the Yosemite Park a wonderland which the Congress of the United States sought by law to reserve for all coming time as nearly as practicable in the condition fashioned by the hand of the Creator—a worthy object of National pride and a source of healthful pleasure and rest for the thousands of people who may annually sojourn there during the heated months.

In 1907 when Mr Garfield became Secretary of the Interior the application was renewed and granted; but under his successor, Mr Fisher, the matter has been referred to a Commission, which as this volume goes to press still has it under consideration.

The most delightful and wonderful camp grounds in the Park are its three great valleys—Yosemite, Hetch Hetchy, and Upper Tuolumne; and they are also the most important places with reference to their positions relative to the other great features—the Merced and Tuolumne Canyons, and the High Sierra peaks and glaciers, etc., at the head of the rivers. The main part of the Tuolumne Valley is a spacious flowery lawn four or five miles long, surrounded by magnificent snowy mountains, slightly separated from other beautiful meadows, which together make a series about twelve miles in length, the highest reaching to the feet of Mount Dana, Mount Gibbs, Mount Lyell and Mount McClure. It is about 8500 feet above the sea, and forms the grand central High Sierra camp ground from which excursions are made to the noble mountains, domes, glaciers, etc,; across the Range to the Mono Lake and volcanoes and down the Tuolumne Canyon to Hetch Hetchy. Should Hetch Hetchy be submerged for a reservoir, as proposed, not only would it be utterly destroyed, but the sublime canyon way to the heart of the High Sierra would be hopelessly blocked and the great camping ground, as watershed of a city drinking system, virtually would be closed to the public. So far as I have learned, few of all the thousands who have seen the park and seek rest and peace in it are in favour of this outrageous scheme.

One of my later visits to the Valley was made in the autumn of 1907 with the late William Keith, the artist. The leaf-colours were then ripe, and the treat godlike rocks in repose seemed to glow with life. The artist, under their spell, wandered day after day along the river and through the groves and gardens, studying the wonderful scenery; and, after making about forty sketches, declared with enthusiasm that although its walls were less sublime in height, in picturesque beauty and charm Hetch Hetchy surpassed even Yosemite.

That any one would try to destroy such a place seems incredible; but sad experience shows that there are people good enough and bad enough for anything. The proponents of the dam scheme bring forward a lot of bad arguments to prove that the only righteous thing to do with the people's parks is to destroy them bit by bit as they are able. Their arguments are curiously like those of the devil, devised for the destruction of the first garden—so much of the very best Eden fruit going to waste; so much of the best Tuolumne water and Tuolumne scenery going to waste. Few of their statements are even partly true, and all are misleading.

Thus, Hetch Hetchy, they say, is a "low-lying meadow." On the contrary, it is a high-lying natural landscape garden.

"It is a common minor feature, like thousands of others." On the contrary it is a very uncommon feature; after Yosemite, the rarest and in many ways the most important in the National Park.

"Damming and submerging it 175 ft. deep would enhance its beauty by forming a crystal-clear lake." Landscape gardens, places of recreation and worship, are never made beautiful by destroying and burying them. The beautiful sham lake, forsooth, would be only an eyesore, a dismal blot on the landscape, like many others to be seen in the Sierra. For, instead of keeping it at the same level all the year, allowing Nature centuries of time to make new shores, it would, of course, be full only a month or two in the spring, when the snow is melting fast; then it would be gradually drained, exposing the slimy sides of the basin and shallower parts of the bottom, with the gathered drift and waste, death and decay of the upper basins, caught here instead of being swept on to decent natural burial along the banks of the river or in the sea. Thus the Hetch Hetchy dam-lake would be only a rough imitation of a natural lake for a few of the spring months, an open sepulchre for the others.

"Hetch Hetchy water is the purest of all to be found in the Sierra, unpolluted, and forever unpollutable." On the contrary, excepting that of the Merced below Yosemite, it is less pure than that of most of the other Sierra streams, because of the sewerage of camp grounds draining into it, especially of the Big Tuolumne Meadows camp ground, occupied by hundreds of tourists and mountainers, with their animals, for months every summer, soon to be followed by thousands from all the world.

These temple destroyers, devotees of ravaging commercialism, seem to have a perfect contempt for Nature, and, instead of lifting their eyes to the God of the mountains, lift them to the Almighty Dollar.

Dam Hetch Hetchy! As well as dam for water-tanks the people's cathedrals and churches, for no holier temple has ever been consecrated by the heart of man.

14

John Burroughs (1837–1921)

The seventh of ten children, John Burroughs was born on a small dairy farm in the foothills of New York's Catskill Mountains on April 3, 1837. Just after his seventeenth birthday he began a career as a country schoolteacher in nearby Ulster County. During the summer months Burroughs would pursue his own education, and he soon came under the influence of Emerson's writings, which he later said, "were like the sunlight to my pale and tender genius which had fed on Johnson and Addison and poor Whipple." In October 1863, Burroughs moved to Washington D.C., where he soon obtained an appointment to the Department of the Treasury, and formed a close friendship with Walt Whitman. For the next three decades Burroughs was an important supporter of Whitman and his poetry and his first book (partly ghostwritten by Whitman himself) was about the poet. In 1871 Burroughs published his first book of natural history essays, Wake-Robin, *which received laudatory reviews and launched his career as America's most popular nature writer. Over the next fifty years, Burroughs published thirty books as well as hundreds of essays in many of the country's leading periodicals. He built a writing cabin that he named Slabsides near his Hudson Valley farm, where he hosted numerous friends and admirers, including Theodore Roosevelt, Henry Ford, and John Muir. In 1903 he was the guest of President Roosevelt on a tour of Yellowstone Park, an experience that Burroughs described in* Camping and Tramping with Roosevelt *(1907). Roosevelt dedicated his own* Outdoor Pastimes of an American Hunter *(1905) to Burroughs, writing: "It is a good thing for our people that you have lived, and surely no man can wish to have more said of him."*

Burroughs was sometimes criticized for his reluctance to use his tremendous popularity as a nature writer to greater effect in the political battle over conservation and wilderness preservation, as did his friend John Muir. Despite his aversion to political activity, Burroughs's contribution to the popularity of nature study in the late nineteenth and earlier twentieth century should not be underestimated. Not only did he attract millions of readers to nature, but the popularity of his books helped bring readers to other nature writers such as Thoreau and Muir, and later, to the support of causes such as conservation, bird protection, and wilderness preservation. Additionally, Burroughs rejected anthropocentrism and raised significant questions about the role that religion had in separating humankind from nature. In essays such as "The Faith of a Naturalist" Burroughs argued that amid the "decay of creeds," the love of nature fostered a spirituality that was not at odds with human reason. As Bill McKibben points out in a 1992 essay, while Burroughs's ideas may not seem to be revolutionary, in many ways "we are still groping toward them."

"The Faith of a Naturalist" from *Accepting the Universe* (1920)

I

To say that man is as good as God would to most persons seem like blasphemy; but to say that man is as good as Nature would disturb no one. Man is a part of Nature, or a phase of Nature, and shares in what we call her imperfections. But what is Nature a part of, or a phase of?—and what or who is its author? Is it not true that this earth which is so familiar to us is as good as yonder morning or evening star and made of the same staff?—just as much in the heavens, just as truly a celestial abode as it is? Venus seems to us like a great jewel in the crown of night or morning. From Venus the earth would seem like a still larger jewel. The heavens seem afar off and free from all stains and impurities of earth; we lift our eyes and our hearts to them as to the face of the Eternal, but our science reveals no body or place there so suitable for human abode and human happiness as this earth. In fact, this planet is the only desirable heaven of which we have any clue. Innumerable other worlds exist in the abysses of space which may be the abodes of beings superior, and of beings inferior, to ourselves. We place our gods afar off so as to dehumanize them, never suspecting that when we do so we discount their divinity. The more human we are,—remembering that to err is human,—the nearer God we are. Of course good and bad are human concepts and are a verdict upon created things as they stand related to us, promoting or hindering our well-being. In the councils of the Eternal there is apparently no such distinction.

Man is not only as good as God; some men are a good deal better, that is, from our point of view; they attain a degree of excellence of which there is no hint in nature—moral excellence. It is not until we treat man as a part of nature—as a product of the earth as literally as are the trees—that we can reconcile these contradictions. If we could build up a composite man out of all the peoples of the earth, including even the Prussians, he would represent fairly well the God in nature.

Communing with God is communing with our own hearts, our own best selves, not with something foreign and accidental. Saints and devotees have gone into the wilderness to find God; of course they took God with them, and the silence and detachment enabled them to hear the still, small voice of their own souls, as one hears the ticking of his own watch in the stillness of the night. We are not cut off, we are not isolated points; the great currents flow through us and over us and around us, and unite us to the whole of nature. Moses saw God in the burning bush, saw him with the eyes of early man whose divinities were clothed in the extraordinary, the fearful, or the terrible; we see him in the meanest weed that grows, and hear him in the gentle murmur of our own heart's blood. The language of devotion and religious conviction is only the language of soberness and truth written large and aflame with emotion.

Man goes away from home searching for the gods he carries with him always. Man can know and feel and love only man. There *is* a deal of sound psychology in the new religion called Christian Science—in that part which emphasizes the power of the mind over the body, and the fact that the world is largely what we make it, that evil is only the shadow of good—old truths reburnished. This helps us to understand the hold it has taken upon such a large number of admirable persons. Good and evil are relative terms, but evil is only the shadow of good. Disease is a reality, but not in the same sense that health is a reality. Positive and negative electricity are both facts, but positive and negative good belong to a different order. Christian Science will not keep the distemper out of the house if the sewer-gas gets in; inoculation will do more to prevent typhoid and diphtheria than "declaring the truth" or saying your prayers or counting your beads. In its therapeutical value experimental science is the only safe guide in dealing with human corporal ailments.

We need not fear alienation from God. I feed Him when I feed a beggar. I serve Him when I serve my neighbor. I love Him when I love my friend. I praise Him when I praise the wise and good of any race or time. I shun Him when I shun the leper. I forgive Him when I forgive my enemies. I wound him when I wound a human being. I forget Him when I forget my duty to others. If I am cruel or unjust or resentful or envious or inhospitable toward any man, woman, or child, I am guilty of all these things toward God: "Inasmuch as ye have done it unto one of the least of these my brethren, ye have done it unto me."

II

I am persuaded that a man without religion falls short of the proper human ideal. Religion, as I use the term, is a spiritual flowering, and the man who has it not is like a plant that never blooms. The mind that does not open and unfold its religious sensibilities in the sunshine of this infinite and spiritual universe, is to be pitied. Men of science do well enough with no other religion than the love of truth, for this is indirectly a love of God. The astronomer, the geologist, the biologist, tracing the footsteps of the Creative Energy throughout the universe—what need has he of any formal, patent-right religion? Were not Darwin, Huxley, Tyndall, and Lyell, and all other seekers and verifiers of natural truth among the most truly religious of men? Any of these men would have gone to hell for the truth—not the truth of creeds and rituals, but the truth as it exists in the councils of the Eternal and as it is written in the laws of matter and of life.

For my part I had a thousand times rather have Huxley's religion than that of the bishops who sought to discredit him, or Bruno's than that of the church that burnt him. The religion of a man that has no other aim than his own personal safety from some real or imaginary future calamity, is of the selfish, ignoble kind.

Amid the decay of creeds, love of nature has high religious value. This has saved many persons in this world—saved them from mammon-worship, and from the frivolity and insincerity of the crowd. It has made their lives placid and sweet. It has given them an inexhaustible field for inquiry, for enjoyment, for the exercise of all their powers, and in the end has not left them soured and dissatisfied. It has made them contented and at home wherever they are in nature—in the house not made with hands. This house is their church, and the rocks and the hills are the altars, and the creed is written in the leaves of the trees and in the flowers of the field and in the sands of the shore. A new creed every day and new preachers, and holy days all the week through. Every walk to the woods is a religious rite, every bath in the stream is a saving ordinance. Communion service is at all hours, and the bread and wine are from the heart and marrow of Mother Earth. There are no heretics in Nature's church; all are believers, all are communicants. The beauty of natural religion is that you have it all the time; you do not have to seek it afar off in myths and legends; in catacombs, in garbled texts, in miracles of dead saints or wine-bibbing friars. It is of to-day; it is now and here; it is everywhere. The crickets chirp it, the birds sing it, the breezes chant it, the thunder proclaims it, the streams murmur it, the unaffected man lives it. Its incense rises from the plowed fields, it is on the morning breeze, it is in the forest breath and in the spray of the wave. The frosts write it in exquisite characters, the dews impearl it, and the rainbow paints it on the cloud. It is not an insurance policy underwritten by a bishop or a priest; it is not even a faith; it is a love, an enthusiasm, a consecration to natural truth.

The God of sunshine and of storms speaks a less equivocal language than the God of revelation.

Our fathers had their religion and their fathers had theirs, but they were not ours, and could not be in those days and under those conditions. But their religions lifted them above themselves; they healed their wounds; they consoled them for many of the failures and disappointments of this world; they developed character; they tempered the steel in their nature. How childish to us seems the plan of salvation, as our fathers found it in the fervid and, I freely say, inspired utterances of Saint Paul! But it saved them, it built character, it made life serious, it was an heroic creed which has lost credence in our more knowing and more frivolous age. We see how impossible it is, but we do not see the great natural truths upon which it rests.

A man is not saved by the truth of the things he believes, but by the truth of his belief—its sincerity, its harmony with his character. The absurdities of the popular religions do not matter; what matters is the lukewarm belief, the empty forms, the shallow conceptions of life and duty. We are prone to think that if the creed is false, the religion is false. Religion is an emotion, an inspiration, a feeling of the Infinite, and may have its root in any creed or in no creed. What can be more unphilosophical than the doctrines of the Christian Scientists? Yet Christian Science is a good practical religion. It makes people cheerful, happy, and helpful—yes, and helps make them healthy too. Its keynote is love, and love holds the universe together. Any creed that ennobles character and opens a door or a window upon the deeper meanings of this marvelous universe is good enough to live by, and good enough to die by. The Japanese-Chinese religion of ancestor worship, sincerely and devoutly held, is better than the veneer of much of our fashionable well-dressed religion.

Guided by appearances alone, how surely we should come to look upon the sun as a mere appendage of the earth!—as much so as is the moon. How near it seems at sunrise and sunset, and as if these phenomena directly involved the sun, extending to it and modifying its light and heat! We do not realize that these are merely terrestrial phenomena, and that the sun, so to speak, knows them not.

Viewed from the sun the earth is a mere speck in the sky, and the amount of the total light and heat from the sun that is received on the earth is so small that the mind can hardly grasp it. Yet for all practical purposes the sun shines for us alone. Our relation to it could not be any more direct and sustaining if it were created for that purpose. It is immanent in the life of the globe. It is the source of all our energy and therefore of our life. Its bounties are universal. The other planets find it is their sun also. It is as special and private to them as to us. We think the sun paints the bow on the cloud, but the bow follows from the laws of optics. The sun knows it not.

It is the same with what we call God. His bounty is of the same universal, impersonal kind, and yet for all practical purposes it exists especially for us, it is immanent every moment in our lives. There is no special Providence. Nature sends the rain upon the just and the unjust, upon the sea as upon the land. We are here and find life good because Providence is general and not special. The conditions are not too easy, the struggle has made men of us. The bitter has tempered the sweet. Evil has put us on our guard and keeps us so. We pay for what we get.

III

That wise old Roman, Marcus Aurelius, says, "Nothing is evil which is according to nature." At that moment he is thinking especially of death which, when it comes in the course of nature, is not an evil, unless life itself is also an evil. After the lamp of life is burned out, death is not an evil, rather is it a good. But premature death, death by accident or disease, before a man has done his work or used up his capital of vitality, is an evil. Disease itself is an evil, but if we lived according to nature there would be no disease; we should die the natural, painless death of old age. Of course there is no such thing as absolute evil or absolute good. Evil is that which is against our well-being, and good is that which promotes it. We always postulate the existence of life when we speak of good and evil. Excesses in nature are evil to us because they bring destruction and death in their train. They are disharmonies in the scheme of things, because they frustrate and bring to naught. The war which Marcus Aurelius was waging when be wrote those passages was an evil in itself, though good might come out of it.

Everything in organic nature—trees, grasses, flowers, insects, fishes, mammals—is beset by evil of some kind. The natural order is good because it brought us here and keeps us here, but evil has always dogged our footsteps. Leaf-blight is an evil to the tree, smallpox is an evil to man, frost is an evil to the insects, flood an evil to the fishes.

Moral evil—hatred, envy, greed, lying, cruelty, cheating—is of another order. These vices have no existence below the human sphere. We call them evils because they are disharmonies; they are inimical to the highest standard of human happiness and well-being. They make a man less a man, they work discord and develop needless friction. Sand in the engine of your car and water in the gasoline are evils, and malice and jealousy and selfishness in your heart are analogous evils.

In our day we read the problem of Nature and God in a new light, the light of science, or of emancipated human reason, and the old myths mean little to us. We accept Nature as we find it, and do not crave the intervention of a God that sits behind and is superior to it. The self-activity of the cosmos suffices. We accept the tornadoes and earthquakes and world wars, and do not lose faith. We arm ourselves against them as best we can. We accept the bounty of the rain, the sunshine, the soil, the changing seasons, and the vast armory of non-living forces, and from them equip or teach ourselves to escape, endure, modify, or ward off the destructive and non-human forces that beset our way. We draw our strength from the Nature that seems and is so regardless of us; our health and wholeness are its gifts. The biologic ages, with all their carnival of huge and monstrous forms, had our well-being at heart. The evils and dangers that beset our way have been outmatched by the good and the helpful. The deep-sea fish would burst and die if brought to the surface; the surface life would be crushed and killed in the deep sea. Life adapts itself to its environment; hard conditions make it hard. Winds, floods, inclement seasons, have driven it around the earth; the severer the cold, the thicker the fur; compensations always abound. If Nature is not all-wise and all-merciful from our human point of view, she has placed us in a world where our own wisdom and mercy can be developed; she has sent us to a school in which we learn to see her own shortcomings and imperfections, and to profit by them.

The unreasoning, unforeseeing animals suffer more from the accidents of nature—drought, flood, lightning—than man does; but man suffers more from evils of his own making—war, greed, intemperance, pestilence—so that the development in both lines goes on, and life is still at the flood.

Good and evil are inseparable. We cannot have light without shade, or warmth without cold, or life without death, or development without struggle. The struggle *for* life, of which Darwinism makes so much, is only the struggle of the chick to get out of the shell, or of the flower to burst its bud, or of the root to penetrate the soil. It is not the struggle of battle and hate—the justification of war and usurpation—it is for the most part a beneficent struggle with the environment, in which the fittest of the individual units of a species survive, but in which the strong and the feeble, the great and the small of species alike survive. The lamb survives with the lion, the wren with the eagle, the Esquimo with the European—all manner of small and delicate forms survive with the great and robust. One species of carnivora, or of rodents, or herbivora, does not, as a rule, exterminate another species. It is true that species prey upon species, that cats eat mice, that hawks eat smaller birds, and that man slays and eats the domestic animals. Probably man alone has exterminated species. But outside of man's doings all the rest belongs to Nature's system of checks and balances, and bears no analogy to human or inhuman wars and conquests.

Life struggles with matter, the tree struggles with the wind and with other trees. Man struggles with gravity, cold, wet, heat, and all the forces that hinder him. The tiniest plant that grows has to force its root down into the soil; earlier than that it has to burst its shell or case. The corn struggles to lift itself up after the storm has beaten it down; effort, effort, everywhere in the organic world. Says Whitman:

> Urge and urge and urge,
> Always the procreant urge of the world.

IV

Every few years we have an ice-storm or a snowstorm that breaks down and disfigures the trees. Some trees suffer much more than others. The storm goes its way; the laws of physical force prevail; the great world of mechanical forces is let loose upon the small world of vital forces; occasionally a tree is so crushed that it never entirely recovers; but after many years the woods and groves have repaired the damages and taken on their wonted thrifty appearance. The evil was only temporary; the world of trees has suffered no permanent setback. But had the trees been conscious beings, what a deal of suffering they would have experienced! An analogous visitation to human communities entails a heritage of misery, but in time it too is forgotten and its scars healed. Fire, blood, war, epidemics, earthquakes, are such visitations, but the race survives them and reaps good from them.

We say that Nature cares nothing for the individual, but only for the race or the species. The whole organic world is at war with the inorganic, and as in human wars the individuals are sacrificed that the army, the whole, may live; so in the strife and competition of nature, the separate units fall that the mass may prosper.

It is probably true that in the course of the biological history of the earth, whole species have been rendered extinct by parasites, or by changing outward conditions. But this has been the exception, and not the rule. The chestnut blight now seems to threaten the very existence of this species of tree in this country, but I think the chances are that this fungus will meet with some natural check.

In early summer comes the June drop of apples. The trees start with more fruit than they can carry, and if they are in vigorous health, they will drop the surplus. It is a striking illustration of Nature's methods. The tree does its own thinning. But if not at the top of its condition, it fails to do this. It takes health and strength simply to let go; only a living tree drops its fruit or its leaves; only a growing man drops his outgrown opinions.

If we put ourselves in the place of the dropped apples, we must look upon our fate as unmixed evil. If we put ourselves in the place of the tree and of the apples that remain on it, the June drop would appear an unmixed good—finer fruit, and a healthier, longer-lived tree results. Nature does not work so much to specific as to universal ends. The individual may go, but the type must remain. The ranks may be decimated, but the army and its cause must triumph. Life in all its forms is a warfare only in the sense that it is a struggle with its outward conditions, in which, other things being equal, the strongest force prevails. Small and weak forms prevail also, because the competing forms are small and weak, or because at the feast of life there is a place for the small and weak also. But lion against lion, man against man, mouse against mouse, the strongest will, in the end, be the victor.

Man's effort is to save waste, to reduce friction, to take short cuts, to make smooth the way, to seize the advantage, to economize time, but the physical forces know none of these things.

Go into the woods and behold the evil the trees have to contend with—all typical of the evil we have to contend with—too crowded in places, one tree crushing another by its fall, specimens on every hand whose term of life might be lengthened by a little wise surgery; borers, blight, disease, insect pests, storm, wreckage, thunderbolt scars, or destruction—evil in a hundred forms besetting every tree, and sooner or later leaving its mark. A few escape—oaks, maples, pines, elms—and reach a greater age than the others, but they fail at last, and when they have rounded out their green century, or ten centuries, and go down in a gale, or in the stillness of a summer night, how often younger trees are marred or crushed by their fall! But come back after many long years, and their places are filled, and all the scars are healed. The new generation of trees is feeding upon the accumulations of the old. Evil is turned to good. The destruction of the cyclone, the ravages of fire, the wreckage of the ice-storm, are all obliterated and the forest-spirit is rank and full again.

There is no wholesale exemption from this rule of waste and struggle in this world, nor probably in any other. We have life on these terms. The organic world develops under pressure from within and from without. Rain brings the perils of rain, fire brings the perils of fire, power brings the perils of power. The great laws go our way, but they will break us or rend us if we fail to keep step with them. Unmixed good is a dream; unmixed happiness is a dream; perfection is a dream; heaven and hell are both dreams of our mixed and struggling lives, the one the outcome of our aspirations for the good, the other the outcome of our fear of evil.

The trees in the woods, the plants in the fields encounter hostile forces the year through; storms crash or overthrow them; visible and invisible enemies prey upon them; yet are the fields clothed in verdure and the hills and plains mantled with superb forests. Nature's haphazard planting and sowing and her wasteful weeding and trimming do not result in failure as these methods do with us. A failure of hers with one form or species results in the success of some other form. All successes are hers. Allow time enough and the forest returns in the path of the tornado, but maybe with other species of trees. The birds and squirrels plant oaks and chestnuts amid the pines and the winds plant pines amid the oaks and chestnuts. The robins and the cedar-birds sow the red cedar broadcast over the landscape, and plant the Virginia creeper and the poison-ivy by every stub and fence-post. The poison-ivy is a triumph of Nature as truly as is the grapevine or the morning-glory. All are hers. Man specializes; he selects this or that, selects the wheat and rejects the tares; but Nature generalizes; she has the artist's disinterestedness; all is good; all are parts of her scheme. She nourishes the foul-smelling catbrier as carefully as she does the rose. Each creature, with man at the head, says, "The world is mine; it was created for me." Evidently it was created for all, at least all forms are at home here. Nature's system of checks and balances preserves her working equilibrium. If a species of forest worm under some exceptionally favoring conditions gets such a start that it threatens to destroy our beech and maple forests, presently a parasite, stimulated by this turn in its favor, appears and restores the balance. For two or three seasons the beechwoods in my native town were ravaged by some kind of worm or beetle; in midsummer the sunlight came into them as if the roof had been taken off; later they swarmed with white millers. But the scourge was suddenly checked—some parasite, probably a species of ichneumon fly, was on hand to curtail the dangerous excess.

I am only trying to say that after we have painted Nature as black as the case will allow, after we have depicted her as a savage beast, a devastating storm, a scorching desert, a consuming fire, an all-engulfing earthquake, or as war, pestilence, famine, we have only depicted her from our limited human point of view. But even from that point of view the favoring conditions of life are so many, living bodies are so adaptive, the lift of the evolutionary impulse is so unconquerable, the elemental laws and forces are so overwhelmingly on our side, that our position in the universe is still an enviable one. "Though he slay me, yet will I trust in him." Slain, I shall nourish some other form of life, and the books will still balance—not my books, but the vast ledgers of the Eternal.

In the old times we accounted for creation in the simple terms of the Hebrew Scriptures—"In the beginning God created the heaven and the earth." We even saw no discrepancy in the tradition that creation took place in the spring. But when we attempt to account for creation in the terms of science or naturalism, the problem is far from being so simple. We have not so tangible a point from which to start. It is as if we were trying to find the end or the beginning of the circle. Round and round we go, caught in the endless and beginningless currents of the Creative Energy; no fixity or finality anywhere; rest and motion, great and small, up and down, heat and cold, good and evil, near and far, only relative; cause and effect merging and losing themselves in each other; life and death perpetually playing into each other's hands; interior within interior; depth beneath depth; height above height; the tangible thrilled and vibrating with the intangible; the material in bonds to the non-material; invisible, impalpable forces streaming around us and through us; perpetual change and transformation on every hand; every day a day of creation, every night a revelation of unspeakable grandeur; suns and systems forming in the cyclones of stardust; the whole starry host of heaven flowing like a meadow brook, but where, or whence, who can tell? The center everywhere, the circumference nowhere; pain and pleasure, good and evil, inextricably mixed; the fall of man a daily and hourly occurrence; the redemption of man, the same! Heaven or hell waiting by every doorstep, boundless, beginningless, unspeakable, immeasurable—what wonder that we seek a short cut through this wilderness and appeal to the supernatural?

When I look forth upon the world and see how, regardless of man and his well-being, the operations of Nature go on—how the winds and the storms wreck him or destroy him, how the drought or the floods bring to naught his industries, how not the least force in heaven or earth turns aside for him, or makes any exception to him; in short, how all forms of life are perpetually ground between the upper and the nether millstones of the contending and clashing natural material forces, I ask myself: "Is there nothing, then, under the sun, or beyond the sun, that has a stake in our well-being? Is life purely a game of chance, and is it all luck that we are here in a world so richly endowed to meet all our requirements?" Serene Reason answers: "No, it is not luck as in a lottery. It is the good fortune of the whole. It was inherent in the constitution of the whole, and it continues because of its adaptability; life is here because it fits itself into the scheme of things; it is flexible and compromising." We find the world good to be in because we are adapted to it, and not it to us. The vegetable growth upon the rocks where the sea is forever pounding is a type of life; the waves favor its development. Life takes advantage of turbulence as well as of quietude, of drought as well as of floods, of deserts as well as of marshes, of the sea-bottom as well as of the mountain-tops. Both animal and vegetable life trim their sails to the forces that beat upon them. The image of the sail is a good one. Life avails itself of the half-contrary winds; it captures and imprisons their push in its sails; by yielding a little, it makes headway in the teeth of the gale; it gives and takes; without struggle, without opposition, life would not be life. The sands of the shore do not struggle with the waves, nor the waves with the sands; the buffeting ends where it began. But trees struggle with the wind, fish struggle with the flood, man struggles with his environment; all draw energy from the forces that oppose them. Life gains as it spends; its waste is an investment. Not so with purely material bodies. They are like the clock, they must be perpetually wound from without. A living body is a clock, perpetually self-wound from within.

The faith and composure of the naturalist or naturist are proof against the worst that Nature can do. He sees the cosmic forces only; he sees nothing directly mindful of man, but man himself; he sees the intelligence and beneficence of the universe flowering in man; he sees life as a mysterious issue of the warring element; he sees human consciousness and our sense of right and wrong, of truth and justice, as arising in the evolutionary sequence, and turning and sitting in judgment upon all things; he sees that there can be no life without pain and death; that there can be no harmony without discord; that opposites go hand in hand; that good and evil are inextricably mingled; that the sun and blue sky are still there behind the clouds, unmindful of them; that all is right with the world if we extend our vision deep enough; that the ways of Nature are the ways of God if we do not make God in our own image, and make our comfort and well-being the prime object of Nature. Our comfort and well-being are provided for in the constitution of the world, but we may say that they are not guaranteed; they are contingent upon many things, but the chances are upon our side. He that would save his life shall lose it—lose it in forgetting that the universe is not a close corporation, or a patented article, and that it exists for other ends than our own. But he who can lose his life in the larger life of the whole shall save it in a deeper, truer sense.

15

"Forever Wild" Provision of the New York State Constitution (Constitutional Convention of 1894)

After years of concern about the effect of deforestation on the Adirondack Mountains, Governor David B. Hill signed into law Chapter 283 of the Laws of 1885, a measure that provided that several hundred thousand acres of state-owned land in the Adirondacks would be preserved as "forever wild." The law was soon discovered to have loopholes large enough to (literally) drive a railroad through, and at New York's Constitutional Convention of 1894, a group of conservationists led by Colonel David McClure, a delegate from New York City, spearheaded a successful effort to insert a true "forever wild" clause into the state constitution that provided protection for the largest forest preserve east of the Mississippi River. While the "Forever Wild" provision has been amended several times to provide for the needs of communities within the boundaries of the Adirondack Park, it endures as one of the most successful and far-reaching attempts by a state to preserve and expand its wilderness regions.

Report of Committee on Forest Preserves

Document No. 63.

To the Constitutional Convention:

The Special Committee on State Forest Preservation, which was directed to consider and report what, if any, amendments to the Constitution should be adopted for the preservation of the forests, respectfully report:

That your committee has had presented to it many valuable arguments and statements bearing upon the matter, and, after careful consideration, has unanimously reached the conclusion that it is necessary for the health, safety, and general advantage of the people of the State that the forest lands now owned, and hereafter acquired by the State and the timber on such lands, should be preserved intact as forest preserves, and not under any circumstances be sold.

Your committee is further of the opinion that, for the perfect protection and preservation of the State lands, other lands contiguous thereto should, as soon as possible, be purchased or otherwise acquired, but feel that any action to that end is more properly within the province of the Legislature than of this Convention.

Your committee recommends the adoption by this Convention of the following as an amendment to the Constitution, viz.:

"The lands of the State now owned, or hereafter acquired, constituting the forest preserves, shall be forever kept as wild forest lands. They shall not, nor shall the timber thereon, be sold."

Dated August 23, 1894.

DAVID McCLURE,
Chairman.

State of New York in Convention. August 23, 1894.

Introduced by special committee on state forest preservation—ordered to third reading September twelfth.

PROPOSED CONSTITUTIONAL AMENDMENT

To amend the constitution relative to the forest preserve.

The Delegates of the People of the State of New York, in Convention assembled, do propose as follows:

ARTICLE—.

Section—.The lands of the state now owned or hereafter acquired, constituting the forest preserve, as now fixed by law, shall be forever kept as wild forest lands. They shall not be leased, sold or exchanged, or be taken by any corporation, public or private, nor shall the timber thereon be sold, removed or destroyed.

New York State Constitution Article 7, Section 7 (As ratified by the Constitutional Convention of 1894)

Section 7. The lands of the State, now owned or hereafter acquired, constituting the forest preserve as now fixed by law, shall be forever kept as wild forest lands. They shall not be leased, sold or exchanged, or be taken by any corporation public or private, nor shall the timber thereon be sold, removed or destroyed.

16

Theodore Roosevelt (1858–1919)

Theodore Roosevelt's vigorous support of conservation programs and wilderness protection was at least in part attributable to his desire to preserve a frontier ethic that had been part of "the pleasantest, healthiest, and most exciting phase of American existence." While Roosevelt's political legacy as a conservationist is often viewed with ambivalence by modern scholars of environmental history, there is little doubt that his presidential administration was the high water mark of the early conservation movement. Working closely with his chief forester Gifford Pinchot, Roosevelt created more national parks, forest preserves, and national monuments than any American president before or since. Roosevelt's numerous books on outdoor life evince an appreciation of nature that goes well beyond the well-established and oft caricatured image of Roosevelt as hunter. Likewise, his friendship with John Burroughs, the nation's most beloved nature writer was not simply political cover for his hunting, but arose in large measure from a shared love of birding. Even John Muir, who openly criticized Roosevelt's hunting and frequently butted heads with Pinchot over conservation issues, seems to have developed a genuine fondness and respect for Roosevelt during a camping trip in Yosemite that the two men took in 1903, writing, "Camping with the President was a remarkable experience. I fairly fell in love with him."

Governor's Annual Message to the State of New York (1900)

Fisheries, Forest, and Game Commission

Under this commission great progress has been made through the fish-hatcheries in the propagation of valuable food and sporting fish. The laws for the protection of deer have resulted in their increase. Nevertheless, as railroads tend to encroach on the wilderness, the temptation to illegal hunting becomes greater, and the danger from forest-fires increases. There is need of great improvement both in our laws and in their administration. The game-wardens have been too few in number. More should be provided. None save fit men must be appointed and their retention in office must depend purely upon the zeal, ability, and efficiency with which they perform their duties. The game-wardens in the forests must be woodsmen; and they should have no outside business. In short, there should be a thorough reorganization of the work of the commission. A careful study of the resources and condition of the forests on State land must be made. It is certainly not too much to expect that the State forests should be managed as efficiently as the forests on private lands in the same neighborhoods, and the measure of difference in efficiency of management must be the measure of condemnation or praise of the way the public forests have been managed.

The subject of forest preservation is of the utmost importance to the State. The Adirondacks and Catskills should be great parks kept in perpetuity for the benefit and enjoyment of our people. Much has been done of late years toward their preservation, but very much remains to be done. The provisions of law in reference to sawmills and wood-pulp mills are defective and should be changed so as to prohibit dumping dyestuff, sawdust, or tan-bark in any amount whatsoever into the streams. Reservoirs should be made; but not where they will tend to destroy large sections of the forest, and only after a careful and scientific study of the water resources of the region. The people of the forest regions are themselves growing more and more to realize the necessity of preserving both the trees and the game. A live deer in the woods will attract to the neighborhood ten times the money that could be obtained for the deer's dead carcass. Timber theft on the State lands is, of course, a grave offense against the whole public.

Hardy outdoor sports, like hunting, are in themselves of no small value to the national character, and should be encouraged in every way. Men who go into the wilderness, indeed men who take part in any field-sports with horse or rifle, receive a benefit which can hardly be given by even the most vigorous athletic games.

There is a further, and more immediate and practical, end in view. A primeval forest is a great sponge which absorbs and distils the rain-water; and when it is destroyed the result is apt to be an alternation of flood and drought. Forest-fires ultimately make the land a desert, and are a detriment to all that portion of the State tributary to the streams through the woods where they occur. Every effort should be made to minimize their destructive influence. We need to have our system of forestry gradually developed and conducted along scientific principles. When this has been done it will be possible to allow marketable lumber to be cut everywhere without damage to the forests—indeed, with positive advantage to them; but until lumbering is thus conducted, on strictly scientific principles no less than upon principles of the strictest honesty toward the State, we cannot afford to suffer it at all in the State forests. Unrestrained greed means the ruin of the great woods and the drying up of the sources of the rivers.

Ultimately the administration of the State lands must be so centralized as to enable us definitely to place responsibility in respect to everything concerning them, and to demand the highest degree of trained intelligence in their use.

The State should not permit within its limits factories to make bird-skins or bird-feathers into articles of ornament or wearing-apparel. Ordinarily birds, and especially song-birds, should be rigidly protected. Game-birds should never be shot to a greater extent than will offset the natural rate of increase. All spring shooting should be prohibited and efforts made by correspondence with the neighboring States to secure its prohibition within their borders. Care should be taken not to encourage the use of cold storage or other market systems which are a benefit to no one but the wealthy epicure who can afford to pay a heavy price for luxuries. These systems tend to the destruction of the game; which would bear most severely upon the very men whose rapacity has been appealed to in order to secure its extermination.

The open season for the different species of game and fish should be made uniform throughout the entire State, save that it should be shorter on Long Island for certain species which are not plentiful, and which are pursued by a greater number of people than in other game portions of the State.

From his First Annual Message as President (December 3, 1901)

FOREST CONSERVATION

Public opinion throughout the United States has moved steadily toward a just appreciation of the value of forests, whether planted or of natural growth. The great part played by them in the creation and maintenance of the national wealth is now more fully realized than ever before.

Wise forest protection does not mean the withdrawal of forest resources, whether of wood, water, or grass, from contributing their full share to the welfare of the people, but, on the contrary, gives the assurance of larger and more certain supplies. The fundamental idea of forestry is the perpetuation of forests by use. Forest protection is not an end of itself; it is a means to increase and sustain the resources of our country and the industries which depend upon them. The preservation of our forests is an imperative business necessity. We have come to see clearly that whatever destroys the forest, except to make way for agriculture threatens our well-being.

The practical usefulness of the national forest reserves to the mining, grazing, irrigation, and other interests of the regions in which the reserves lie has led to a widespread demand by the people of the West for their protection and extension. The forest reserves will inevitably be of still greater use in the future than in the past. Additions should be made to them whenever practicable, and their usefulness should be increased by a thoroughly businesslike management.

At present the protection of the forest reserves rests with the General Land Office, the mapping and description of their timber with the United States Geological Survey, and the preparation of plans for their conservative use with the Bureau of Forestry, which is also charged with the general advancement of practical forestry in the United States. These various functions should be united in the Bureau of Forestry, to which they properly belong. The present diffusion of responsibility is bad from every standpoint. It prevents that effective co-operation between the government and the men who utilize the resources of the reserves, without which the interests of both must suffer. The scientific bureaus generally should be put under the Department of Agriculture. The President should have by law the power of transferring lands for use as forest reserves to the Department of Agriculture. He already has such power in the case of lands needed by the Departments of War and the Navy.

The wise administration of the forest reserves will be not less helpful to the interests which depend on water than to those which depend on wood and grass. The water-supply itself depends upon the forest. In the arid region it is water, not land, which measures production. The western half of the United States would sustain a population greater than that of our whole country today if the waters that now run to waste were saved and used for irrigation. The forest and water problems are perhaps the most vital internal questions of the United States.

Game Protection

Certain of the forest reserves should also be made preserves for the wild forest creatures. All of the reserves should be better protected from fires. Many of them need special protection because of the great injury done by live stock, above all by sheep. The increase in deer, elk, and other animals in the Yellowstone Park shows what may be expected when other mountain forests are properly protected by law and properly guarded. Some of these areas have been so denuded of surface vegetation by overgrazing that the ground-breeding birds, including grouse and quail, and many mammals, including deer, have been exterminated or driven away. At the same time the water-storing capacity of the surface has been decreased or destroyed, thus promoting floods in times of rain and diminishing the flow of streams between rains.

In cases where natural conditions have been restored for a few years, vegetation has again carpeted the ground, birds and deer are coming back, and hundreds of persons, especially from the immediate neighborhood, come each summer to enjoy the privilege of camping. Some at least of the forest reserves should afford perpetual protection to the native fauna and flora, safe havens of refuge to our rapidly diminishing wild animals of the larger kinds, and free camping-grounds for the ever-increasing numbers of men and women who have learned to find rest, health, and recreation in the splendid forests and flower-clad meadows of our mountains. The forest reserves should be set apart forever for the use and benefit of our people as a whole and not sacrificed to the short-sighted greed of a few.

Water Conservation

The forests are natural reservoirs. By restraining the streams in flood and replenishing them in drought they make possible the use of waters otherwise wasted. They prevent the soil from washing, and so protect the storage reservoirs from filling up with silt. Forest conservation is therefore an essential condition of water conservation.

The forests alone cannot, however, fully regulate and conserve the waters of the arid region. Great storage works are necessary to equalize the flow of streams and to save the flood waters. Their construction has been conclusively shown to be an undertaking too vast for private effort. Nor can it be best accomplished by the individual States acting alone. Far-reaching interstate problems are involved; and the resources of single States would often be inadequate. It is properly a national function, at least in some of its features. It is as right for the National Government to make the streams and rivers of the arid region useful by engineering works for water storage as to make useful the rivers and harbors of the humid region by engineering works of another kind. The storing of the floods in reservoirs at the headwaters of our rivers is but an enlargement of our present policy of river control, under which levees are built on the lower reaches of the same streams.

The government should construct and maintain these reservoirs as it does other public works. Where their purpose is to regulate the flow of streams, the water should be turned freely into the channels in the dry season to take the same course under the same laws as the natural flow.

Reclamation and Irrigation

The reclamation of the unsettled arid public lands presents a different problem. Here it is not enough to regulate the flow of streams. The object of the government is to dispose of the land to settlers who will build homes upon it. To accomplish this object water must be brought within their reach.

The pioneer settlers on the arid public domain chose their homes along streams from which they could themselves divert the water to reclaim their holdings. Such opportunities are practically gone. There remain, however, vast areas of public land which can be made available for homestead settlement, but only by reservoirs and mainline canals impracticable for private enterprise. These irrigation works should be built by the National Government. The lands reclaimed by them should be reserved by the government for actual settlers, and the cost of construction should so far as possible be repaid by the land reclaimed. The distribution of the water, the division of the streams among irrigators, should be left to the settlers themselves in conformity with State laws and without interference with those laws or with vested rights. The policy of the National Government should be to aid irrigation in the several States and Territories in such manner as will enable the people in the local communities to help themselves, and as will stimulate needed reforms in the State laws and regulations governing irrigation.

The reclamation and settlement of the arid lands will enrich every portion of our country, just as the settlement of the Ohio and Mississippi valleys brought prosperity to the Atlantic States. The increased demand for manufactured articles will stimulate industrial production, while wider home markets and the trade of Asia will consume the larger food-supplies and effectually prevent Western competition with Eastern agriculture. Indeed, the products of irrigation will be consumed chiefly in upbuilding local centres of mining and other industries, which would otherwise not come into existence at all. Our people as a whole will profit, for successful home-making is but another name for the upbuilding of the nation.

The necessary foundation has already been laid for the inauguration of the policy just described. It would be unwise to begin by doing too much, for a great deal will doubtless be learned, both as to what can and what cannot be safely attempted, by the early efforts, which must of necessity be partly experimental in character. At the very beginning the government should make clear, beyond shadow of doubt, its intention to pursue this policy on lines of the broadest public interest. No reservoir or canal should ever be built to satisfy selfish personal or local interests; but only in accordance with the advice of trained experts, after long investigation has shown the locality where all the conditions combine to make the work most needed and fraught with the greatest usefulness to the community as a whole. There should be no extravagance, and the believers in the need of irrigation will most benefit their cause by seeing to it that it is free from the least taint of excessive or reckless expenditure of the public moneys.

Whatever the nation does for the extension of irrigation should harmonize with, and tend to improve, the condition of those now living on irrigated land. We are not at the starting-point of this development. Over two hundred millions of private capital has already been expended in the Construction of irrigation works, and many million acres of arid land reclaimed. A high degree of enterprise and ability has been shown in the work itself; but as much cannot be said in reference to the laws relating thereto. The security and value of the homes created depend largely on the stability of titles to water; but the majority of these rest on the uncertain foundation of court decisions rendered in ordinary suits at law. With a few creditable exceptions, the arid States have failed to provide for the certain and just division of streams in times of scarcity. Lax and uncertain laws have made it possible to establish rights to water in excess of actual uses or necessities, and many streams have already passed into private ownership, or a control equivalent to ownership.

Whoever controls a stream practically controls the land it renders productive, and the doctrine of private ownership of water apart from land cannot prevail without causing enduring wrong. The recognition of such ownership, which has been permitted to grow up in the arid regions, should give way to a more enlightened and larger recognition of the rights of the public in the control and disposal of the public water-supplies. Laws founded upon conditions obtaining in humid regions, where water is too abundant to justify hoarding it, have no proper application in a dry country.

In the arid States the only right to water which should be recognized is that of use. In irrigation this right should attach to the land reclaimed and be inseparable therefrom. Granting perpetual water-rights to others than users, without compensation to the public, is open to all the objections which apply to giving away perpetual franchises to the public utilities of cities. A few of the Western States have already recognized this, and have incorporated in their constitutions the doctrine of perpetual State ownership of water.

The benefits which have followed the unaided development of the past justify the nation's aid and co-operation in the more difficult and important work yet to be accomplished. Laws so vitally affecting homes as those which control the water-supply will only be effective when they have the sanction of the irrigators; reforms can only be final and satisfactory when they come through the enlightenment of the people most concerned. The larger development which national aid insures should, however, awaken in every arid State the determination to make its irrigation system equal in justice and effectiveness that of any country in the civilized world. Nothing could be more unwise than for isolated communities to continue to learn every thing experimentally, instead of profiting by what is already known elsewhere. We are dealing with a new and momentous question, in the pregnant years while institutions are forming, and what we do will affect not only the present but future generations.

Our aim should be not simply to reclaim the largest area of land and provide homes for the largest number of people, but to create for this new industry the best possible social and industrial conditions; and this requires that we not only understand the existing situation, but avail ourselves of the best experience of the time in the solution of its problems. A careful study should be made, both by the nation and the States, of the irrigation laws and conditions here and abroad. Ultimately it will probably be necessary for the nation to co-operate with the several arid States in proportion as these States by their legislation and administration show themselves fit to receive it.

17

GIFFORD PINCHOT (1865–1946)

While a student at Yale in the 1880s, Gifford Pinchot decided that he would pursue a career as a professional forester. Because no such programs were offered by any American universities at the time, he attended the French National School of Forestry. Following his return to the United States in 1890, he took a position with the recently formed United States Bureau of Forestry. He soon left government service to open his own consulting firm, where he developed a reputation as one of the country's leading foresters and conservationists. In 1896 Pinchot served on a forestry commission appointed by Congress to study ways to implement the Forest Reserve. Pinchot was a proponent of resource management or "wise use," and while serving on the commission had a falling out with fellow commission member John Muir over the issue of opening the newly created forest preserves for mining and grazing. Over the next twenty years, this conflict between wilderness preservation and resource utilization would have significant repercussions for the conservation movement, which split into two antagonistic factions over this issue. After Theodore Roosevelt assumed the presidency in 1901, Roosevelt appointed Pinchot as his chief forester, a position he held until he was dismissed by President William Taft in 1910. Pinchot later served two terms (1923–1926 and 1931–1934) as governor of Pennsylvania.

The Fight for Conservation *(1910) reflects both Pinchot's firmly held belief in the Progressive political agenda and his core principles regarding conservation. Unlike Muir and his allies in the wilderness preservation movement, Pinchot believed that natural resources should be used as part of a well-planned program of economic development. Such a program should eliminate unnecessary waste of such resources and ensure that the benefits of their development should be fairly distributed rather than used for the economic benefit of only a few. As Pinchot wrote in the following essay, "Conservation means the greatest good to the greatest number for the longest time. One of its great contributions is just this, that it has added to the worn and well-known phrase, 'the greatest good to the greatest number,' the additional words 'for the longest time,' thus recognizing that this nation of ours must be made to endure as the best possible home for all its people."*

"Principles of Conservation" from *The Fight for Conservation* (1910)

The principles which the word Conservation has come to embody are not many, and they are exceedingly simple. I have had occasion to say a good many times that no other great movement has ever achieved such progress in so short a time, or made itself felt in so many directions with such vigor and effectiveness, as the movement for the conservation of natural resources.

Forestry made good its position in the United States before the conservation movement was born. As a forester I am glad to believe that conservation began with forestry, and that the principles which govern the Forest Service in particular and forestry in general are also the ideas that control conservation.

The first idea of real foresight in connection with natural resources arose in connection with the forest. From it sprang the movement which gathered impetus until it culminated in the great Convention of Governors at Washington in May, 1908. Then came the second official meeting of the National Conservation movement, December, 1908, in Washington. Afterward came the various gatherings of citizens in convention, come together to express their judgment on what ought to be done, and to contribute, as only such meetings can, to the formation of effective public opinion.

The movement so begun and so prosecuted has gathered immense swing and impetus. In 1907 few knew what Conservation meant. Now it has become a household word. While at first Conservation was supposed to apply only to forests, we see now that its sweep extends even beyond the natural resources.

The principles which govern the conservation movement, like all great and effective things, are simple and easily understood. Yet it is often hard to make the simple, easy, and direct facts about a movement of this kind known to the people generally.

The first great fact about conservation is that it stands for development. There has been a fundamental misconception that conservation means nothing but the husbanding of resources for future generations. There could be no more serious mistake. Conservation does mean provision for the future, but it means also and first of all the recognition of the right of the present generation to the fullest necessary use of all the resources with which this country is so abundantly blessed. Conservation demands the welfare of this generation first, and afterward the welfare of the generations to follow.

The first principle of conservation is development, the use of the natural resources now existing on this continent for the benefit of the people who live here now. There may be just as much waste in neglecting the development and use of certain natural resources as there is in their destruction. We have a limited supply of coal, and only a limited supply. Whether it is to last for a hundred or a hundred and fifty or a thousand years, the coal is limited in amount, unless through geological changes which we shall not live to see, there will never be any more of it than there is now. But coal is in a sense the vital essence of our civilization. If it can be preserved, if the life of the mines can be extended, if by preventing waste there can be more coal left in this country after we of this generation have made every needed use of this source of power, then we shall have deserved well of our descendants.

Conservation stands emphatically for the development and use of water-power now, without delay. It stands for the immediate construction of navigable waterways under a broad and comprehensive plan as assistants to the railroads. More coal and more iron are required to move a ton of freight by rail than by water, three to one. In every case and in every direction the conservation movement has development for its first principle, and at the very beginning of its work. The development of our natural resources and the fullest use of them for the present generation is the first duty of this generation. So much for development.

In the second place conservation stands for the prevention of waste. There has come gradually in this country an understanding that waste is not a good thing and that the attack on waste is an industrial necessity. I recall very well indeed how, in the early days of forest fires, they were considered simply and solely as acts of God, against which any opposition was hopeless and any attempt to control them not merely hopeless but childish. It was assumed that they came in the natural order of things, as inevitably as the seasons or the rising and setting of the sun. To-day we understand that forest fires are wholly within the control of men. So we are coming in like manner to understand that the prevention of waste in all other directions is a simple matter of good business. The first duty of the human race is to control the earth it lives upon.

We are in a position more and more completely to say how far the waste and destruction of natural resources are to be allowed to go on and where they are to stop. It is curious that the effort to stop waste, like the effort to stop forest fires, has often been considered as a matter controlled wholly by economic law. I think there could be no greater mistake. Forest fires were allowed to burn long after the people had means to stop them. The idea that men were helpless in the face of them held long after the time had passed when the means of control were fully within our reach. It was the old story that "as a man thinketh, so is he"; we came to see that we could stop forest fires, and we found that the means had long been at hand. When at length we came to see that the control of logging in certain directions was profitable, we found it had long been possible. In all these matters of waste of natural resources, the education of the people to understand that they can stop the leakage comes before the actual stopping and after the means of stopping it have long been ready at our hands.

In addition to the principles of development and preservation of our resources there is a third principle. It is this: The natural resources must be developed and preserved for the benefit of the many, and not merely for the profit of a few. We are coming to understand in this country that public action for public benefit has a very much wider field to cover and a much larger part to play than was the case when there were resources enough for every one, and before certain constitutional provisions had given so tremendously strong a position to vested rights and property in general.

A few years ago President Hadley, of Yale, wrote an article which has not attracted the attention it should. The point of it was that by reason of the XIVth amendment to the Constitution, property rights in the United States occupy a stronger position than in any other country in the civilized world. It becomes then a matter of multiplied importance, since property rights once granted are so strongly entrenched, to see that they shall be so granted that the people shall get their fair share of the benefit which comes from the development of the resources which belong to us all. The time to do that is now. By so doing we shall avoid the difficulties and conflicts which will surely arise if we allow vested rights to accrue outside the possibility of governmental and popular control.

The conservation idea covers a wider range than the field of natural resources alone. Conservation means the greatest good to the greatest number for the longest time. One of its great contributions is just this, that it has added to the worn and well-known phrase, "the greatest good to the greatest number," the additional words "for the longest time," thus recognizing that this nation of ours must be made to endure as the best possible home for all its people.

Conservation advocates the use of foresight, prudence, thrift, and intelligence in dealing with public matters, for the same reasons and in the same way that we each use foresight, prudence, thrift, and intelligence in dealing with our own private affairs. It proclaims the right and duty of the people to act for the benefit of the people. Conservation demands the application of common-sense to the common problems for the common good.

The principles of conservation thus described—development, preservation, the common good—have a general application which is growing rapidly wider. The development of resources and the prevention of waste and loss, the protection of the public interests, by foresight, prudence, and the ordinary business and home-making virtues, all these apply to other things as well as to the natural resources. There is, in fact, no interest of the people to which the principles of conservation do not apply.

The conservation point of view is valuable in the education of our people as well as in forestry; it applies to the body politic as well as to the earth and its minerals. A municipal franchise is as properly within its sphere as a franchise for water-power. The same point of view governs in both. It applies as much to the subject of good roads as to waterways, and the training of our people in citizenship is as germane to it as the productiveness of the earth. The application of common-sense to any problem for the Nation's good will lead directly to national efficiency wherever applied. In other words, and that is the burden of the message, we are coming to see the logical and inevitable outcome that these principles, which arose in forestry and have their bloom in the conservation of natural resources, will have their fruit in the increase and promotion of national efficiency along other lines of national life.

The outgrowth of conservation, the inevitable result, is national efficiency. In the great commercial struggle between nations which is eventually to determine the welfare of all, national efficiency will be the deciding factor. So from every point of view conservation is a good thing for the American people.

The National Forest Service, one of the chief agencies of the conservation movement, is trying to be useful to the people of this nation. The Service recognizes, and recognizes it more and more strongly all the time, that whatever it has done or is doing has just one object, and that object is the welfare of the plain American citizen. Unless the Forest Service has served the people, and is able to contribute to their welfare it has failed in its work and should be abolished. But just so far as by cooperation, by intelligence, by attention to the work laid upon it, it contributes to the welfare of our citizens, it is a good thing and should be allowed to go on with its work.

The Natural Forests are in the West. Headquarters of the Service have been established throughout the Western country, because its work cannot be done effectively and properly without the closest contact and the most hearty cooperation with the Western people. It is the duty of the Forest Service to see to it that the timber, water-powers, mines, and every other resource of the forests is used for the benefit of the people who live in the neighborhood or who may have a share in the welfare of each locality. It is equally its duty to cooperate with all our people in every section of our land to serve a fundamental resource, without which this Nation cannot prosper.

18

Robert Marshall (1901–1939)

Although he was just thirty-eight years old when he died, Bob Marshall is a major figure in the wilderness preservation movement. Marshall's interest in the wilderness began early—his father was a progressive New York lawyer and conservationist, and the family spent their summers in the Adirondacks, where Marshall and his brothers became avid hikers and canoeists. Marshall studied forestry in college, and earned a doctorate in plant pathology from Johns Hopkins University in 1930. As an outdoorsman, Marshall's stamina was legendary; he often hiked 30–40 miles in a day, and in 1932 climbed fourteen different Adirondack peaks in one nineteen-hour period. Following a series of journeys through Alaska's Brooks Range (1929–1931), Marshall's prodigious energy was increasingly directed toward wilderness preservation through his position as director of forestry in the Interior Department's Office of Indian Affairs and later as a cofounder of the Wilderness Society in 1935.

"The Problem of the Wilderness," first published in the February 1930 issue of The Scientific Monthly, *is one of the clearest and most concise arguments ever written in support of wilderness preservation. In this essay Marshall argues that the true value of wilderness cannot be measured in economic terms alone, that or many it also has a physical, mental, and esthetic value, concluding that, "there is just one hope of repulsing the tyrannical ambition of civilization to conquer every niche on the whole earth. That hope is the organization of spirited people who will fight for the freedom of the wilderness." Interestingly, in making his argument for wilderness preservation, Marshall turns around one of the key tenets of Theodore Roosevelt's notion of the "strenuous life," citing Bertrand Russell's assertion that "many men would cease to desire war if they had opportunities to risk their lives in Alpine climbing."*

"The Problem of the Wilderness" (1930)

I

It is appalling to reflect how much useless energy has been expended in arguments which would have been inconceivable had the terminology been defined. In order to avoid such futile controversy I shall undertake at the start to delimit the meaning of the principal term with which this paper is concerned. According to Dr. Johnson a *wilderness* is "a tract of solitude and savageness," a definition more poetic than explicit. Modern lexicographers do better with "a tract of land, whether a forest or a wide barren plain, uncultivated and uninhabited by human beings."[1] This definition gives a rather good foundation, but it still leaves a penumbra of partially shaded connotation.

For the ensuing discussion I shall use the word *wilderness* to denote a region which contains no permanent inhabitants, possesses no possibility of conveyance by any mechanical means and is sufficiently spacious that a person in crossing it must have the experience of sleeping out. The dominant attributes of such an area are: first, that it requires any one who exists in it to depend exclusively on his own effort for survival; and second, that it preserves as nearly as possible the primitive environment. This means that all roads, power transportation and settlements are barred. But trails and temporary shelters, which were common long before the advent of the white race, are entirely permissible.

When Columbus effected his immortal debarkation, he touched upon a wilderness which embraced virtually a hemisphere. The philosophy that progress is proportional to the amount of alteration imposed upon nature never seemed to have occurred to the Indians. Even such tribes as the Incas, Aztecs and Pueblos made few changes in the environment in which they were born. "The land and all that it bore they treated with consideration; not attempting to improve it, they never desecrated it."[2] Consequently, over billions of acres the aboriginal wanderers still spun out their peripatetic careers, the wild animals still browsed in unmolested meadows and the forests still grew and moldered and grew again precisely as they had done for undeterminable centuries.

It was not until the settlement of Jamestown in 1607 that there appeared the germ for that unabated disruption of natural conditions which has characterized all subsequent American history. At first expansion was very slow. The most intrepid seldom advanced further from their neighbors than the next drainage. At the time of the Revolution the zone of civilization was still practically confined to a narrow belt lying between the Atlantic Ocean and the Appalachian valleys. But a quarter of a century later, when the

Reprinted with permission from *The Scientific Monthly*, Vol. 30, No. 2 (February 1930), 141–148. Copyright ©1930, American Association for the Advancement of Science (AAAS).

Louisiana Purchase was consummated, the outposts of civilization had reached the Mississippi, and there were foci of colonization in half a dozen localities west of the Appalachians, though the unbroken line of the frontier was east of the mountains.[3]

It was yet possible as recently as 1804 and 1805 for the Lewis and Clark Expedition to cross two thirds of a continent without seeing any culture more advanced than that of the Middle Stone Age. The only routes of travel were the uncharted rivers and the almost impassable Indian trails. And continually the expedition was breaking upon some "truly magnificent and sublimely grand object, which has from the commencement of time been concealed from the view of civilized man."[4]

This exploration inaugurated a century of constantly accelerating emigration such as the world had never known. Throughout this frenzied period the only serious thought ever devoted to the wilderness was how it might be demolished. To the pioneers pushing westward it was an enemy of diabolical cruelty and danger, standing as the great obstacle to industry and development. Since these seemed to constitute the essentials for felicity, the obvious step was to excoriate the devil which interfered. And so the path of empire proceeded to substitute for the undisturbed seclusion of nature the conquering accomplishments of man. Highways wound up valleys which had known only the footsteps of the wild animals; neatly planted gardens and orchards replaced the tangled confusion of the primeval forest; factories belched up great clouds of smoke where for centuries trees had transpired toward the sky, and the ground-cover of fresh sorrel and twin-flower was transformed to asphalt spotted with chewing-gum, coal dust and gasoline.

To-day there remain less than twenty wilderness areas of a million acres, and annually even these shrunken remnants of an undefiled continent are being despoiled. Aldo Leopold has truly said:

> The day is almost upon us when canoe travel will consist in paddling up the noisy wake of a motor launch and portaging through the back yard of a summer cottage. When that day comes canoe travel will be dead, and dead too will be a part of our Americanism . . . . The day is almost upon us when a pack train must wind its way up a graveled highway and turn out its bell mare in the pasture of a summer hotel. When that day comes the pack train will be dead, the diamond hitch will be merely a rope and Kit Carson and Jim Bridger will be names in a history lesson.[5]

Within the next few years the fate of the wilderness must be decided. This is a problem to be settled by deliberate rationality and not by personal prejudice. Fundamentally, the question is one of balancing the total happiness which will be obtainable if the few un-desecrated areas are perpetuated against that which will prevail if they are destroyed. For this purpose it will be necessary: first, to consider the extraordinary benefits of the wilderness; second, to enumerate the drawbacks to undeveloped areas; third, to evaluate the relative importance of these conflicting factors, and finally, to formulate a plan of action.

II

The benefits which accrue from the wilderness may be separated into three broad divisions: the physical, the mental and the esthetic.

Most obvious in the first category is the contribution which the wilderness makes to health. This involves something more than pure air and quiet, which are also attainable in almost any rural situation. But toting a fifty-pound pack over an abominable trail, snowshoeing across a blizzard-swept plateau or scaling some jagged pinnacle which juts far above timber all develop a body distinguished by a soundness, stamina and élan unknown amid normal surroundings.

More than mere heartiness is the character of physical independence which can be nurtured only away from the coddling of civilization. In a true wilderness if a person is not qualified to satisfy all the requirements of existence, then he is bound to perish. As long as we prize individuality and competence it is imperative to provide the opportunity for complete self-sufficiency. This is inconceivable under the effete superstructure of urbanity; it demands the harsh environment of untrammeled expanses.

Closely allied is the longing for physical exploration which bursts through all the chains with which society fetters it. Thus we find Lindbergh, Amundsen, Byrd gaily daring the unknown, partly to increase knowledge, but largely to satisfy the craving for adventure. Adventure, whether physical or mental, implies breaking into unpenetrated ground, venturing beyond the boundary of normal aptitude, extending oneself to the limit of capacity, courageously facing peril. Life without the chance for such exertions would be for many persons a dreary game, scarcely bearable in its horrible banality.

It is true that certain people of great erudition "come inevitably to feel that if life has any value at all, then that value comes in thought,"[6] and so they regard mere physical pleasures as puerile inconsequences. But there are others, perfectly capable of comprehending relativity and the quantum theory, who find equal ecstacy in non-intellectual adventure. It is entirely irrelevant which view-point is correct; each is applicable to whoever entertains it. The important consideration is that both groups are entitled to indulge their penchant, and in the second instance this is scarcely possible without the freedom of the wilderness.

III

One of the greatest advantages of the wilderness is its incentive to independent cogitation. This is partly a reflection of physical stimulation, but more inherently due to the fact that original ideas require an objectivity and perspective seldom possible in the distracting propinquity of one's fellow men. It is necessary to "have gone behind the world of humanity, seen its institutions like toadstools by the wayside."[7] This theorizing is justified empirically by the number of America's most virile minds, including Thomas Jefferson, Henry Thoreau, Louis Agassiz, Herman Melville, Mark Twain,

John Muir and William James, who have felt the compulsion of periodical retirements into the solitudes. Withdrawn from the contaminating notions of their neighbors, these thinkers have been able to meditate, unprejudiced by the immuring civilization.

Another mental value of an opposite sort is concerned not with incitement but with repose. In a civilization which requires most lives to be passed amid inordinate dissonance, pressure and intrusion, the chance of retiring now and then to the quietude and privacy of sylvan haunts becomes for some people a psychic necessity. It is only the possibility of convalescing in the wilderness which saves them from being destroyed by the terrible neural tension of modern existence.

There is also a psychological bearing of the wilderness which affects, in contrast to the minority who find it indispensable for relaxation, the whole of human kind. One of the most profound discoveries of psychology has been the demonstration of the terrific harm caused by suppressed desires. To most of mankind a very powerful desire is the appetite for adventure. But in an age of machinery only the extremely fortunate have any occasion to satiate this hankering, except vicariously. As a result people become so choked by the monotony of their lives that they are readily amenable to the suggestion of any lurid diversion. Especially in battle, they imagine, will be found the glorious romance of futile dreams. And so they endorse war with enthusiasm and march away to stirring music, only to find their adventure a chimera, and the whole world miserable. It is all tragically ridiculous, and yet there is a passion there which can not be dismissed with a contemptuous reference to childish quixotism. William James has said that "militarism is the great preserver of ideals of hardihood, and human life with no use for hardihood would be contemptible."[8] The problem, as he points out, is to find a "moral equivalent of war," a peaceful stimulation for the hardihood and competence instigated in bloodshed. This equivalent may be realized if we make available to every one the harmless excitement of the wilderness. Bertrand Russell has skilfully amplified this idea in his essay on "Machines and the Emotions." He expresses the significant conclusion that "many men would cease to desire war if they had opportunities to risk their lives in Alpine climbing."[9]

IV

In examining the esthetic importance of the wilderness I will not engage in the unprofitable task of evaluating the preciousness of different sorts of beauty, as, for instance, whether an acronical view over the Grand Canyon is worth more than the Apollo of Praxiteles. For such a rating would always have to be based on a subjective standard, whereas the essential for any measure is impersonality. Instead of such useless metaphysics I shall call attention to several respects in which the undisputed beauty of the primeval, whatever its relative merit, is distinctly unique.

Of the myriad manifestations of beauty, only natural phenomena like the wilderness are detached from all temporal relationship. All the beauties in the creation or alteration of which man has played even the slightest rôle are firmly anchored in the historic stream. They are temples of Egypt, oratory of Rome, painting of the Renaissance or music of the Classicists. But in the wild places nothing is moored more closely than to geologic ages. The silent wanderer crawling up the rocky shore of the turbulent river could be a savage from some prehistoric epoch or a fugitive from twentieth century mechanization.

The sheer stupendousness of the wilderness gives it a quality of intangibility which is unknown in ordinary manifestations of ocular beauty. These are always very definite two or three dimensional objects which can be physically grasped and circumscribed in a few moments. But "the beauty that shimmers in the yellow afternoons of October, who ever could clutch it."[10] Any one who has looked across a ghostly valley at midnight, when moonlight makes a formless silver unity out of the drifting fog, knows how impossible it often is in nature to distinguish mass from hallucination. Any one who has stood upon a lofty summit and gazed over an inchoate tangle of deep canyons and cragged mountains, of sunlit lakelets and black expanses of forest, has become aware of a certain giddy sensation that there are no distances, no measures, simply unrelated matter rising and falling without any analogy to the banal geometry of breadth, thickness and height. A fourth dimension of immensity is added which makes the location of some dim elevation outlined against the sunset as incommensurable to the figures of the topographer as life itself is to the quantitative table of elements which the analytic chemist proclaims to constitute vitality.

Because of its size the wilderness also has a physical ambiency about it which most forms of beauty lack. One looks from outside at works of art and architecture, listens from outside to music or poetry. But when one looks at and listens to the wilderness he is encompassed by his experience of beauty, lives in the midst of his esthetic universe.

A fourth peculiarity about the wilderness is that it exhibits a dynamic beauty. A Beethoven symphony or a Shakespearean drama, a landscape by Corot or a Gothic cathedral, once they are finished become virtually static. But the wilderness is in constant flux. A seed germinates, and a stunted seedling battles for decades against the dense shade of the virgin forest. Then some ancient tree blows down and the long-suppressed plant suddenly enters into the full vigor of delayed youth, grows rapidly from sapling to maturity, declines into the conky senility of many centuries, dropping millions of seeds to start a new forest upon the rotting débris of its own ancestors, and eventually topples over to admit the sunlight which ripens another woodland generation.

Another singular aspect of the wilderness is that it gratifies every one of the senses. There is unanimity in venerating the sights and sounds of the forest. But what are generally esteemed to be the minor senses should not be slighted. No one who has ever strolled in springtime through seas of blooming violets, or lain at night on boughs of fresh balsam, or walked across dank holms in early morning can omit odor from the joys of the primordial environment. No one who has felt the stiff wind of mountaintops or the softness of untrodden sphagnum will forget the exhilaration experienced through touch. "Nothing ever tastes as good as when it's cooked in the woods" is a trite tribute to another sense. Even equilibrium causes a blithe exultation during many a river crossing on tenuous foot log and many a perilous conquest of precipice.

Finally, it is well to reflect that the wilderness furnishes perhaps the best opportunity for pure esthetic enjoyment. This requires that beauty be observed as a unity, and that for the brief duration of any pure esthetic experience the cognition of the observed object must completely fill the spectator's cosmos. There can be no extraneous thoughts—no question about the creator of the phenomenon, its structure, what it resembles or what vanity in the beholder it gratifies. "The purely esthetic observer has for the moment forgotten his own soul";[11] he has only one sensation left and that is exquisiteness. In the wilderness, with its entire freedom from the manifestations of human will, that perfect objectivity which is essential for pure esthetic rapture can probably be achieved more readily than among any other forms of beauty.

V

But the problem is not all one-sided. Having discussed the tremendous benefits of the wilderness, it is now proper to ponder upon the disadvantages which uninhabited territory entails.

In the first place, there is the immoderate danger that a wilderness without developments for fire protection will sooner or later go up in smoke and down in ashes.

A second drawback is concerned with the direct economic loss. By locking up wilderness areas we as much as remove from the earth all the lumber, minerals, range land, water-power and agricultural possibilities which they contain. In the face of the tremendous demand for these resources it seems unpardonable to many to render nugatory this potential material wealth.

A third difficulty inherent in undeveloped districts is that they automatically preclude the bulk of the population from enjoying them. For it is admitted that at present only a minority of the genus *Homo* cares for wilderness recreation, and only a fraction of this minority possesses the requisite virility for the indulgence of this desire. Far more people can enjoy the woods by automobile. Far more would prefer to spend their vacations in luxurious summer hotels set on well-groomed lawns than in leaky, fly-infested shelters bundled away in the brush. Why then should this majority have to give up its rights?

VI

As a result of these last considerations the irreplaceable values of the wilderness are generally ignored, and a fatalistic attitude is adopted in regard to the ultimate disappearance of all unmolested localities. It is my contention that this outlook is entirely unjustified, and that almost all the disadvantages of the wilderness can be minimized by forethought and some compromise.

The problem of protection dictates the elimination of undeveloped areas of great fire hazard. Furthermore, certain infringements on the concept of an unsullied wilderness will be unavoidable in almost all instances. Trails, telephone lines and lookout cabins will have to be constructed, for without such precaution most forests in the west would be gutted. But even with these improvements the basic primitive quality still exists: dependence on personal effort for survival.

Economic loss could be greatly reduced by reserving inaccessible and unproductive terrain. Inasmuch as most of the highly valuable lands have already been exploited, it should be easy to confine a great share of the wilderness tracts to those lofty mountain regions where the possibility of material profit is unimportant. Under these circumstances it seems like the grossest illogicality for any one to object to the withdrawal of a few million acres of low-grade timber for recreational purposes when one hundred million acres of potential forest lie devastated.[12] If one tenth portion of this denuded land were put to its maximum productivity, it could grow more wood than all the proposed wilderness areas put together. Or if our forests, instead of attaining only 22 per cent, of their possible production,[13] were made to yield up to capacity, we could refrain from using three quarters of the timber in the country and still be better off than we are to-day. The way to meet our commercial demands is not to thwart legitimate divertisement, but to eliminate the unmitigated evils of fire and destructive logging. It is time we appreciated that the real economic problem is to see how little land need be employed for timber production, so that the remainder of the forest may be devoted to those other vital uses incompatible with industrial exploitation.

Even if there should be an underproduction of timber, it is well to recall that it is much cheaper to import lumber for industry than to export people for pastime. The freight rate from Siberia is not nearly as high as the passenger rate to Switzerland.

What small financial loss ultimately results from the establishment of wilderness areas must be accepted as a fair price to pay for their unassessable preciousness. We spend about twenty-one billion dollars a year for entertainment of all sorts.[14] Compared with this there is no significance to the forfeiture of a couple of million dollars of annual income, which is all that our maximum wilderness requirements would involve. Think what an enormously greater sum New York City alone sacrifices in the maintenance of Central Park.

But the automobilists argue that a wilderness domain precludes the huge majority of recreation-seekers from deriving any amusement whatever from it. This is almost as irrational as contending that because more people enjoy bathing than art exhibits therefore we should change our picture galleries into swimming pools. It is undeniable that the automobilist has more roads than he can cover in a lifetime. There are upward of 3,000,000[15] miles of public highways in the United States, traversing many of the finest scenic features in the nation. Nor would the votaries of the wilderness object to the construction of as many more miles in the vicinity of the old roads, where they would not be molesting the few remaining vestiges of the primeval. But when the motorists also demand for their particular diversion the insignificant wilderness residue, it makes even a Midas appear philanthropic.

> Such are the differences among human beings in their sources of pleasure, that unless there is a corresponding diversity in their modes of life, they neither obtain their fair share of happiness, nor grow up to the mental, moral and esthetic stature of which their nature is capable. Why then should tolerance extend only to tastes and modes of life which extort acquiescence by the multitude of their adherents?[16]

It is of the utmost importance to concede the right of happiness also to people who find their delight in unaccustomed ways. This prerogative is valid even though its exercise may encroach slightly on the fun of the majority, for there is a point where an increase in the joy of the many causes a decrease in the joy of the few out of all proportion to the gain of the former. This has been fully recognized not only by such philosophers of democracy as Paine, Jefferson and Mill, but also in the practical administration of governments which spend prodigious sums of money to satisfy the expensive wants of only a fragment of the community. Public funds which could bring small additional happiness to the mobility are diverted to support museums, art galleries, concerts, botanical gardens, menageries and golf-links. While these, like wilderness areas, are open to the use of every one, they are vital to only a fraction of the entire population. Nevertheless, they are almost universally approved, and the appropriations to maintain them are growing phenomenally.

VII

These steps of reasoning lead up to the conclusion that the preservation of a few samples of undeveloped territory is one of the most clamant issues before us today. Just a few years more of hesitation and the only trace of that wilderness which has exerted such a fundamental influence in molding American character will lie in the musty pages of pioneer books and the mumbled memories of tottering antiquarians. To avoid this catastrophe demands immediate action.

A step in the right direction has already been initiated by the National Conference on Outdoor Recreation,[17] which has proposed twenty-one possible wilderness areas. Several of these have already been set aside in a tentative way by the Forest Service; others are undergoing more careful scrutiny. But this only represents the incipiency of what ought to be done.

A thorough study should forthwith be undertaken to determine the probable wilderness needs of the country. Of course, no precise reckoning could be attempted, but a radical calculation would be feasible. It ought to be radical for three reasons: because it is easy to convert a natural area to industrial or motor usage, impossible to do the reverse; because the population which covets wilderness recreation is rapidly enlarging and because the higher standard of living which may be anticipated should give millions the economic power to gratify what is to-day merely a pathetic yearning. Once the estimate is formulated, immediate steps should be taken to establish enough tracts to insure every one who hungers for it a generous opportunity of enjoying wilderness isolation.

To carry out this program it is exigent that all friends of the wilderness ideal should unite. If they do not present the urgency of their view-point the other side will certainly capture popular support. Then it will only be a few years until the last escape from society will be barricaded. If that day arrives there will be countless souls born to live in strangulation, countless human beings who will be crushed under the artificial edifice raised by man. There is just one hope of repulsing the tyrannical ambition of civilization to conquer every niche on the whole earth. That hope is the organization of spirited people who will fight for the freedom of the wilderness.

NOTES

1. Webster's New International Dictionary.
2. Willa Cather, "Death Comes for the Archbishop."
3. Frederic L. Paxson "History of the American Frontier."
4. Reuben G. Thwaites, "Original Journals of the Lewis and Clark Expedition, 1804–1806," June 13, 1805.
5. Aldo Leopold, "The Last Stand of the Wilderness," *American Forests and Forest Life*, October, 1925.
6. Joseph Wood Krutch, "The Modern Temper."
7. Henry David Thoreau, "Journals," April 2, 1852.
8. William James, "The Moral Equivalent of War."
9. Bertrand Russell, "Essays in Scepticism."
10. Ralph Waldo Emerson, "Nature."
11. Irwin Edman, "The World, the Arts and the Artist."
12. George P. Ahern, "Deforested America," Washington, D. C.
13. U. S. Department of Agriculture, "Timber, Mine or Crop?"
14. Stuart Chase, "Whither Mankind?"
15. "The World Almanac," 1929.
16. John Stuart Mill, "On Liberty."
17. National Conference on Outdoor Recreation, "Recreation Resources of Federal Lands," Washington, D.C.

19

Aldo Leopold (1887–1948)

In 1900, Gifford Pinchot and Henry S. Graves founded the Yale School of Forestry, where Leopold began his studies to become a professional forester in 1906. At Yale, Leopold was indoctrinated in the resource management principles of forestry espoused by Pinchot. Following his graduation in 1909 Leopold took a position with the U.S. Forest Service, and was assigned to the Apache National Forest in Arizona, although he soon developed a dislike for serving as a "tie-pickler or timber tester" for loggers. Over the next fifteen years, Leopold held a series of Forest Service positions, primarily in the Southwest, where he also wrote and lectured widely on conservation, honing the scientific and rhetorical expertise that is so evident in his later writings on conservation. In 1933, Leopold published Game Management, *an authoritative textbook on wildlife conservation and then accepted a teaching position at the University of Wisconsin's newly established department of game management, a position he held up to his death. Two years later, Leopold became one of the founders of the Wilderness Society, and purchased an abandoned farm in one of the "sand counties" of Wisconsin. At this farm, Leopold put his principles of conservation and ecology into practice, as he and his family worked to reestablish an ecosystem that had been devastated by inefficient farming practices and drought. On April 14, 1948 Leopold died from a heart attack suffered while helping neighbors battle a prairie fire on a nearby farm.*

Like many of his early contemporaries in the field of wildlife conservation, Leopold had been taught to believe that the ruthless extermination of predators would result in a "hunter's paradise." In the 1920s, this notion was dramatically debunked on the Kaibab plateau in Arizona when predator elimination led initially to an explosion in the deer population and then a massive die-off due to overgrazing. This experience had a profound effect on Leopold's ecological philosophy and led to essays such as "Threatened Species," and "The Land Ethic." Essays such as these clearly show how far Leopold had progressed from the utilitarian emphasis of the early conservation movement. In "Threatened Species" Leopold argued that it is not merely game species that should be protected by conservationists, but wildlife—even those species that do not have a readily apparent economic value. Most importantly, he argues that habitat protection is an integral component to protecting endangered species. Nearly forty years later, this notion had permeated the political mainstream and resulted in the passage of the Endangered Species Act (1973).

"Threatened Species" (1936)

The volume of effort expended on wildlife conservation shows a large and sudden increase. This effort originates from diverse courses, and flows through diverse channels toward diverse ends. There is a widespread realization that it lacks coordination and focus.

Government is attempting to secure coordination and focus through reorganization of departments, laws, and appropriations. Citizen groups are attempting the same thing through reorganization of associations and private funds.

But the easiest and most obvious means to coordination has been overlooked: explicit definition of the immediate needs of particular species In particular places. For example: Scores of millions are being spent for land purchase, C.C.C. labor, fences, roads, trails, planting, predator-control, erosion control, poisoning, investigations, water developments, silviculture, irrigation, nurseries, wilderness areas, power dams, and refuges, within the natural range of the grizzly bear.

Few would question the assertion that to perpetuate the grizzly as a part of our national fauna is a prime duty of the conservation movement. Few would question the assertion that any one of these undertakings, at any time and place, may vitally affect the restoration of the grizzly, and make it either easy or impossible of accomplishment. Yet no one has made a list of the specific needs of the grizzly, in each and every spot where he survives, and in each and every spot where he might be reintroduced, so that conservation projects in or near that spot may be judged in the light of whether they *help or hinder* the perpetuation of the noblest of American mammals.

On the contrary, our plans, departments, bureaus, associations, and movements are all focused on abstract categories such as recreation, forestry, parks, nature education, wildlife research, more game, fire control, marsh restoration. Nobody cares anything for these except as means toward ends. What ends? There are of course many ends which cannot and many others which need not be precisely defined at this time. But it admits of no doubt that the immediate needs of threatened members of our fauna and flora must be defined now or not at all.

Until they are defined and made public, we cannot blame public agencies, or even private ones, for misdirected effort, crossed wires, or lost opportunities. It must not be forgotten that the abstract categories we have set up as

Reprinted with permission of the University of Wisconsin Press, from *The River of the Mother of God and Other Essays by Aldo Leopold*, Susan L. Flader and J. Baird Callicott, eds. Copyright © 1991.

conservation objectives may serve as alibis for blunders, as well as ends for worthy work. I cite in evidence the C.C.C. crew which chopped down one of the few remaining eagle's nests in northern Wisconsin, in the name of "timber stand improvement." To be sure, the tree was dead, and according to the rules, constituted a fire risk.

Most species of shootable non-migratory game have at least a fighting chance of being saved through the process of purposeful manipulation of laws and environment called management. However great the blunders, delays, and confusion in getting management of game species under way, it remains true that powerful motives of local self-interest are at work in their behalf. European countries, through the operation of these motives, have saved their resident game. It is an ecological probability that we will evolve ways to do so.

The same cannot be said, however, of those species of wilderness game which do not adapt themselves to economic land-use, or of migratory birds which are owned in common, or of non-game forms classed as predators, or of rare plant associations which must compete with economic plants and livestock, or in general of all wild native forms which fly at large or have only an esthetic and scientific value to man. These, then, are the special and immediate concern of this inventory. Like game, these forms depend for their perpetuation on protection and a favorable environment. They need "management"—the perpetuation of good habitat—just as game does, but the ordinary motives for providing it are lacking. They are the threatened element in outdoor America—the crux of conservation policy. The new organizations which have now assumed the name "wildlife" instead of "game," and which aspire to implement the wildlife movement, are I think obligated to focus a substantial part of their effort on these threatened forms.

This is a proposal, not only for an inventory of threatened forms in each of their respective places of survival, but an inventory of the information, techniques, and devices applicable to each species in each place, and of local human agencies capable of applying them. Much information exists, but it is scattered in many minds and documents. Many agencies are or would be willing to use it, if it were laid under their noses. If for a given problem no information exists, or no agency exists, that in itself is useful inventory.

For example, certain ornithologists have discovered a remnant of the Ivory-billed Woodpecker—a bird inextricably interwoven with our pioneer tradition—the very spirit of that "dark and bloody ground" which has become the locus of the national culture. It is known that the Ivory-bill requires as its habitat large stretches of virgin hardwood. The present remnant lives in such a forest, owned and held by an industry as reserve stumpage. Cutting may begin, and the Ivory-bill may be done for at any moment. The Park Service has or can get funds to buy virgin forests, but it does not know of the Ivory-bill or its predicament. It is absorbed in the intricate problem of accommodating the public which is mobbing its parks. When it buys a new park, it is likely to do so in some "scenic" spot, with the general objective of making room for more visitors, rather than with the specific objective of perpetuating some definite thing to visit, its wildlife program is befogged with the abstract concept of inviolate sanctuary. Is it not time to establish particular parks or their equivalent for particular "natural wonders" like the Ivory-bill?

You may say, of course, that one rare bird is no park project—that the Biological Survey should buy a refuge, or the Forest Service a National Forest, to take care of the situation. Whereupon the question bounces back: the Survey has only duck money; the Forest Service would have to cut the timber. But is there anything to prevent the three possible agencies concerned from getting together and agreeing whose job this is, and while they are at it, a thousand other jobs of like character? And how much each would cost? And just what needs to be done in each case? And can anyone doubt that the public, through Congress, would support such a program? Well—this is what I mean by an inventory and plan.

Some sample lists of the items which need to be covered are wilderness and other game species, such as grizzly bear, desert and bighorn sheep, caribou, Minnesota remnants of spruce partridge, masked bobwhite, Sonora deer, peccary, sagehen; predator and allied species, such as the wolf, fisher, otter, wolverine and Condor; migratory birds, including the trumpeter swan, curlews, sandhill crane, Brewster's warbler; plant associations, such as prairie floras, bog floras, Alpine and swamp floras.

In addition to these forms, which are rare everywhere, there is the equally important problem of preserving the attenuated edges of species common at their respective centres. The turkey in Colorado, or the ruffed grouse in Missouri, or the antelope in Nebraska, are rare species within the meaning of this document. That there are grizzlies in Alaska is no excuse for letting the species disappear from New Mexico.

It is important that the inventory represent not merely a protest of those privileged to think, but an agreement of those empowered to act. This means that the inventory should be made by a joint committee of the conservation bureaus, plus representatives of the Wildlife Conference as representing the states and the associations. The plan for each species should be a joint commitment of what is to be done and who is to do it. The bureaus, with their avalanche of appropriations, ought to be able to loan the necessary expert personnel for such a committee, without extra cost. To sift out any possible imputation of bureaucratic, financial, or clique interest, the inter-bureau committee should feed its findings to the public through a suitable group in the National Research Council, and subject to the Council's approval. The necessary incidental funds for a secretary, for expense of gathering testimony and maps, and for publications might well come from the Wildlife Institute, or from one of the scientific foundations.

There is one cog lacking in the hoped-for machine: a means to get some kind of responsible care of remnants of wildlife remote from any bureau or its field officers. Funds can hardly be found to set up special paid personnel for each such detached remnant. It is of course proved long ago that closed seasons and refuge posters without personnel are of no avail. Here is where associations with their far-flung chapters, state officers or departments, or even private individuals can come to the rescue. One of the tragedies of contemporary conservation is the isolated individual or group who complains of having no job. The lack is not of jobs, but of eyes to see them.

The inventory should be the conservationist's eye. Every remnant should be definitely entrusted to a custodian—ranger, warden, game manager, chapter, ornithologist, farmer, stockman, lumberjack. Every conservation meeting—national, state, or local—should occupy itself with hearing their annual reports. Every field inspector should contact their custodians—he might often learn as well as teach. I am satisfied that thousands of enthusiastic conservationists would be proud of such a public trust, and many would execute it with fidelity and intelligence.

I can see in this set-up more conservation than could be bought with millions of new dollars, more coordination of bureaus than Congress can get by new organization charts, more genuine contacts between factions than will ever occur in the war of the inkpots, more research than would accrue from many gifts, and more public education than would accrue from an army of orators and organizers. It is, in effect, a vehicle for putting Jay Darling's concept of "ancestral ranges" into action on a quicker and wider scale than could be done by appropriations alone.

"The Land Ethic" from *A Sand Country Almanac* (1949)

In "The Land Ethic" (1949), Leopold proposes both a new ecological ethic ("A thing is right when it tends to preserve the integrity, stability, and beauty of the biotic community. It is wrong when it tends otherwise"), and a radical change in the anthropocentric way in which we see the world, arguing that "a land ethic changes the role of Homo Sapiens *from conqueror of the land-community to plain member and citizen of it. It implies respect for his fellow-members, and also respect for the community as such." Like earlier conservationists, Leopold generally focused on land use issues in his work, but the biocentric ethical system proposed in "The Land Ethic" continues to influence environmental discourse.*

When God-like Odysseus returned from the wars in Troy, he hanged all on one rope a dozen slave-girls of his household whom he suspected of misbehavior during his absence.

This hanging involved no question of propriety. The girls were property. The disposal of property was then, as now, a matter of expediency, not of right and wrong.

Concepts of right and wrong were not lacking from Odysseus' Greece: witness the fidelity of his wife through the long years before at last his

Reprinted with permission of Oxford University Press, Inc., "The Land Ethic," from *A Sand County Almanac and Sketches Here and There* by Aldo Leopold. Copyright © 1949, 1953, 1966, renewed 1977, 1981 by Oxford University Press, Inc.

black-prowed galleys clove the wine-dark seas for home. The ethical structure of that day covered wives, but had not yet been extended to human chattels. During the three thousand years which have since elapsed, ethical criteria have been extended to many fields of conduct, with corresponding shrinkages in those judged by expediency only.

The Ethical Sequence

This extension of ethics, so far studied only by -philosophers, is actually a process in ecological evolution. Its sequences may be described in ecological as well as in philosophical terms. An ethic, ecologically, is a limitation on freedom of action in the struggle for existence. An ethic, philosophically, is a differentiation of social from anti-social conduct. These are two definitions of one thing. The thing has its origin in the tendency of interdependent individuals or groups to evolve modes of co-operation. The ecologist calls these symbioses. Politics and economics are advanced symbioses in which the original free-for-all competition has been replaced, in part, by co-operative mechanisms with an ethical content.

The complexity of co-operative mechanisms has increased with population density, and with the efficiency of tools. It was simpler, for example, to define the anti-social uses of sticks and stones in the days of the mastodons than of bullets and billboards in the age of motors.

The first ethics dealt with the relation between individuals; the Mosaic Decalogue is an example. Later accretions dealt with the relation between the individual and society. The Golden Rule tries to integrate the individual to society; democracy to integrate social organization to the individual.

There is as yet no ethic dealing with man's relation to land and to the animals and plants which grow upon it. Land, like Odysseus' slave-girls, is still property. The land-relation is still strictly economic, entailing privileges but not obligations.

The Land Ethic

The extension of ethics to this third element in human environment is, if I read the evidence correctly, an evolutionary possibility and an ecological necessity. It is the third step in a sequence. The first two have already been taken. Individual thinkers since the days of Ezekiel and Isaiah have asserted that the despoliation of land is not only inexpedient but wrong. Society, however, has not yet affirmed their belief. I regard the present conservation movement as the embryo of such an affirmation.

An ethic may be regarded as a mode of guidance for meeting ecological situations so new or intricate, or involving such deferred reactions, that the path of social expediency is not discernible to the average individual. Animal instincts are modes of guidance for the individual in meeting such situations. Ethics are possibly a kind of community instinct in-the-making.

The Community Concept

All ethics so far evolved rest upon a single premise: that the individual is a member of a community of interdependent parts. His instincts prompt him to compete for his place in the community, but his ethics prompt him also to co-operate (perhaps in order that there may be a place to compete for).

The land ethic simply enlarges the boundaries of the community to include soils, waters, plants, and animals, or collectively: the land.

This sounds simple: do we not already sing our love for and obligation to the land of the free and the home of the brave? Yes, but just what and whom do we love? Certainly not the soil, which we are sending helter-skelter downriver. Certainly not the waters which we assume have no function except to turn turbines, float barges, and carry off sewage. Certainly not the plants, of which we exterminate whole communities without batting an eye. Certainly not the animals, of which we have already extirpated many of the largest and most beautiful species. A land ethic of course cannot prevent the alteration, management, and use of these "resources," but it does affirm their right to continued existence, and, at least in spots, their continued existence in a natural state.

In short, a land ethic changes the role of *Homo sapiens* from conqueror of the land-community to plain member and citizen of it. It implies respect for his fellow-members, and also respect for the community as such.

In human history, we have learned (I hope) that the conqueror role is eventually self-defeating. Why? Because it is implicit in such a role that the conqueror knows, *ex cathedra*, just what makes the community clock tick, and just what and who is valuable, and what and who is worthless, in community life. It always turns out that he knows neither, and this is why his conquests eventually defeat themselves.

In the biotic community, a parallel situation exists. Abraham knew exactly what the land was for: it was to drip milk and honey into Abraham's mouth. At the present moment, the assurance with which we regard this assumption is inverse to the degree of our education.

The ordinary citizen today assumes that science knows what makes the community clock tick; the scientist is equally sure that he does not. He knows that the biotic mechanism is so complex that its workings may never be fully understood.

That man is, in fact, only a member of a biotic team is shown by an ecological interpretation of history. Many historical events, hitherto explained solely in terms of human enterprise, were actually biotic interactions between people and land. The characteristics of the land determined the facts quite as potently as the characteristics of the men who lived on it.

Consider, for example, the settlement of the Mississippi valley. In the years following the Revolution, three groups were contending for its control: the native Indian, the French and English traders, and the American settlers. Historians wonder what would have happened if the English at Detroit had thrown a little more weight into the Indian side of those tipsy scales which decided the outcome of the colonial migration into the cane-lands of Kentucky. It is time now to ponder the fact that the cane-lands, when subjected to the particular mixture of forces represented by the cow, plow, fire, and axe of the pioneer, became bluegrass. What if the plant succession inherent in this dark and bloody ground had, under the impact of these forces, given us some worthless sedge, shrub, or weed? Would Boone and Kenton have held out? Would there have been any overflow into Ohio, Indiana, Illinois, and Missouri? Any Louisiana Purchase? Any transcontinental union of new states? Any Civil War?

Kentucky was one sentence in the drama of history. We are commonly told what the human actors in this drama tried to do, but we are seldom told that their success, or the lack of it, hung in large degree on the reaction of particular soils to the impact of the particular forces exerted by their occupancy. In the case of Kentucky, we do not even know where the bluegrass came from—whether it is a native species, or a stowaway from Europe.

Contrast the cane-lands with what hindsight tells us about the Southwest, where the pioneers were equally brave, resourceful, and persevering. The impact of occupancy here brought no bluegrass, or other plant fitted to withstand the bumps and buffetings of hard use. This region, when grazed by livestock, reverted through a series of more and more worthless grasses, shrubs, and weeds to a condition of unstable equilibrium. Each recession of plant types bred erosion; each increment to erosion bred a further recession of plants. The result today is a progressive and mutual deterioration, not only of plants and soils, but of the animal community subsisting thereon. The early settlers did not expect this: on the ciénegas of New Mexico some even cut ditches to hasten it. So subtle has been its progress that few residents of the region are aware of it. It is quite invisible to the tourist who finds this wrecked landscape colorful and charming (as indeed it is, but it bears scant resemblance to what it was in 1848).

This same landscape was "developed" once before, but with quite different results. The Pueblo Indians settled the Southwest in pre-Columbian times, but they happened *not* to be equipped with range live-stock. Their civilization expired, but not because their land expired.

In India, regions devoid of any sod-forming grass have been settled, apparently without wrecking the land, by the simple expedient of carrying the grass to the cow, rather than vice versa. (Was this the result of some deep wisdom, or was it just good luck? I do not know.)

In short, the plant succession steered the course of history; the pioneer simply demonstrated, for good or ill, what successions inhered in the land. Is history taught in this spirit? It will be, once the concept of land as a community really penetrates our intellectual life.

The Ecological Conscience

Conservation is a state of harmony between men and land. Despite nearly a century of propaganda, conservation still proceeds at a snail's pace; progress still consists largely of letterhead pieties and convention oratory. On the back forty we still slip two steps backward for each forward stride.

The answer to this dilemma is "more conservation education." No one will debate this, but is it certain that only the *volume* of education needs stepping up? Is something lacking in the *content* as well?

It is difficult to give a fair summary of its content in brief form, but, as I understand it, the content is substantially this: obey the law, vote right, join some organizations, and practice what conservation is profitable on your own land; the government will do the rest.

Is not this formula too easy to accomplish anything worth-while? It defines no right or wrong, assigns no obligation, calls for no sacrifice, implies no change in the current philosophy of values. In respect of land-use, it urges only enlightened self-interest. Just how far will such education take us? An example will perhaps yield a partial answer.

By 1930 it had become clear to all except the ecologically blind that southwestern Wisconsin s topsoil was slipping seaward. In 1933 the farmers were told that if they would adopt certain remedial practices for five years, the public would donate CCC labor to install them, plus the necessary machinery and materials. The offer was widely accepted, but the practices were widely forgotten when the five-year contract period was up. The farmers continued only those practices that yielded an immediate and visible economic gain for themselves.

This led to the idea that maybe farmers would learn more quickly if they themselves wrote the rules. Accordingly the Wisconsin Legislature in 1937 passed the Soil Conservation District Law. This said to farmers, in effect: *We, the public, will furnish you free technical service and loan you specialized machinery, if you will write your own rules for land-use. Each county may write its own rules, and these will have the force of* law. Nearly all the counties promptly organized to accept the proffered help, but after a decade of operation, *no county has yet written a single rule.* There has been visible progress in such practices as strip-cropping, pasture renovation, and soil liming, but none in fencing woodlots against grazing, and none in excluding plow and cow from steep slopes. The farmers, in short, have selected those remedial practices which were profitable anyhow, and ignored those which were profitable to the community, but not clearly profitable to themselves.

When one asks why no rules have been written, one is told that the community is not yet ready to support them; education must precede rules. But the education actually in progress makes no mention of obligations to land over and above those dictated by self-interest. The net result is that we have more education but less soil, fewer healthy woods, and as many floods as in 1937.

The puzzling aspect of such situations is that the existence of obligations over and above self-interest is taken for granted in such rural community enterprises as the betterment of roads, schools, churches, and baseball teams. Their existence is not taken for granted, nor as yet seriously discussed, in bettering the behavior of the water that falls on the land, or in the preserving of the beauty or diversity of the farm landscape. Land-use ethics are still governed wholly by economic self-interest, just as social ethics were a century ago.

To sum up: we asked the farmer to do what he conveniently could to save his soil, and he has done just that, and only that. The farmer who clears the woods off a 75 per cent slope, turns his cows into the clearing, and dumps its rainfall, rocks, and soil into the community creek, is still (if otherwise decent) a respected member of society. If he puts lime on his fields and plants his crops on contour, he is still entitled to all the privileges and emoluments of his Soil Conservation District. The District is a beautiful piece of social machinery, but it is coughing along on two cylinders because we have been too timid, and too anxious for quick success, to tell the farmer the true magnitude of his obligations. Obligations have no meaning without conscience, and the problem we face is the extension of the social conscience from people to land.

No important change in ethics was ever accomplished without an internal change in our intellectual emphasis, loyalties, affections, and convictions. The proof that conservation has not yet touched these foundations of conduct lies in the fact that philosophy and religion have not yet heard of it. In our attempt to make conservation easy, we have made it trivial.

SUBSTITUTES FOR A LAND ETHIC

When the logic of history hungers for bread and we hand out a stone, we are at pains to explain how much the stone resembles bread. I now describe some of the stones which serve in lieu of a land ethic.

One basic weakness in a conservation system based wholly on economic motives is that most members of the land community have no economic value. Wildflowers and songbirds are examples. Of the 22,000 higher plants and animals native to Wisconsin, it is doubtful whether more than 5 per cent can be sold, fed, eaten, or otherwise put to economic use. Yet these creatures are members of the biotic community and if (as I believe) its stability depends on its integrity, they are entitled to continuance.

When one of these non-economic categories is threatened, and if we happen to love it, we invent subterfuges to give it economic importance. At the beginning of the century songbirds were supposed to be disappearing. Ornithologists jumped to the rescue with some distinctly shaky evidence to the effect that insects would eat us up if birds failed to control them. The evidence had to be economic in order to be valid.

It is painful to read these circumlocutions today. We have no land ethic yet, but we have at least drawn nearer the point of admitting that birds should continue as a matter of biotic right, regardless of the presence or absence of economic advantage to us.

A parallel situation exists in respect of predatory mammals, raptorial birds, and fish-eating birds. Time was when biologists somewhat overworked the evidence that these creatures preserve the health of game by killing weaklings, or that they control rodents for the farmer, or that they prey only on "worthless" species. Here again, the evidence had to be economic in order to be valid. It is only in recent years that we hear the more honest argument that predators are members of the community, and that no special interest has the right to exterminate them for the sake of a benefit, real or fancied, to itself. Unfortunately this enlightened view is still in the talk stage. In the field the extermination of predators goes merrily on: witness the impending erasure of the timber wolf by fiat of Congress, the Conservation Bureaus, and many state legislatures.

Some species of trees have been "read out of the party" by economics-minded foresters because they grow too slowly, or have too low a sale value to pay as timber crops: white cedar, tamarack, cypress, beech, and hemlock are examples. In Europe, where forestry is ecologically more advanced, the non-commercial tree species are recognized as members of the native forest community, to be preserved as such, within reason. Moreover some (like beech) have been found to have a valuable function in building up soil fertility. The interdependence of the forest and its constituent tree species, ground flora, and fauna is taken for granted.

Lack of economic value is sometimes a character not only of species or groups, but of entire biotic communities: marshes, bogs, dunes, and "deserts" are examples. Our formula in such cases is to relegate their conservation to government as refuges, monuments, or parks. The difficulty is that these communities are usually interspersed with more valuable private lands; the government cannot possibly own or control such scattered parcels. The net effect is that we have relegated some of them to ultimate extinction over large areas. If the private owner were ecologically minded, he would be proud to be the custodian of a reasonable proportion of such areas, which add diversity and beauty to his farm and to his community.

In some instances, the assumed lack of profit in these "waste" areas has proved to be wrong, but only after most of them had been done away with. The present scramble to reflood muskrat marshes is a case in point.

There is a clear tendency in American conservation to relegate to government all necessary jobs that private landowners fail to perform. Government ownership, operation, subsidy, or regulation is now widely prevalent in forestry, range management, soil and watershed management, park and wilderness conservation, fisheries management, and migratory bird management, with more to come. Most of this growth in governmental conservation is proper and logical, some of it is inevitable. That I imply no disapproval of it is implicit in the fact that I have spent most of my life working for it. Nevertheless the question arises: What is the ultimate magnitude of the enterprise? Will the tax base carry its eventual ramifications? At what point will governmental conservation, like the mastodon, become handicapped by its own dimensions? The answer, if there is any, seems to be in a land ethic, or some other force which assigns more obligation to the private landowner.

Industrial landowners and users, especially lumbermen and stockmen, are inclined to wail long and loudly about the extension of government ownership and regulation to land, but (with notable exceptions) they show little disposition to develop the only visible alternative: the voluntary practice of conservation on their own lands.

When the private landowner is asked to perform some unprofitable act for the good of the community, he today assents only with outstretched palm. If the act costs him cash this is fair and proper, but when it costs only fore-thought, open-mindedness, or time, the issue is at least debatable. The overwhelming growth of land-use subsidies in recent years must be ascribed, in large part, to the government's own agencies for conservation education: the land bureaus, the agricultural colleges, and the extension services. As far as I can detect, no ethical obligation toward land is taught in these institutions.

To sum up: a system of conservation based solely on economic self-interest is hopelessly lopsided. It tends to ignore, and thus eventually to eliminate, many elements in the land community that lack commercial value, but that are (as far as we know) essential to its healthy functioning. It assumes, falsely, I think, that the economic parts of the biotic clock will function without the uneconomic parts. It tends to relegate to government many functions eventually too large, too complex, or too widely dispersed to be performed by government.

An ethical obligation on the part of the private owner is the only visible remedy for these situations.

The Land Pyramid

An ethic to supplement and guide the economic relation to land presupposes the existence of some mental image of land as a biotic mechanism. We can be ethical only in relation to something we can see, feel, understand, love, or otherwise have faith in.

The image commonly employed in conservation education is "the balance of nature." For reasons too lengthy to detail here, this figure of speech fails to describe accurately what little we know about the land mechanism. A much truer image is the one employed in ecology: the biotic pyramid. I shall first sketch the pyramid as a symbol of land, and later develop some of its implications in terms of land-use.

Plants absorb energy from the sun. This energy flows through a circuit called the biota, which may be represented by a pyramid consisting of layers. The bottom layer is the soil. A plant layer rests on the soil, an insect layer on the plants, a bird and rodent layer on the insects, and so on up through various animal groups to the apex layer, which consists of the larger carnivores.

The species of a layer are alike not in where they came from, or in what they look like, but rather in what they eat. Each successive layer depends on those below it for food and often for other services, and each in turn furnishes food and services to those above. Proceeding upward, each successive layer decreases in numerical abundance. Thus, for every carnivore there are hundreds of his prey, thousands of their prey, millions of insects, uncountable plants. The pyramidal form of the system reflects this numerical progression from apex to base. Man shares an intermediate layer with the bears, raccoons, and squirrels which eat both meat and vegetables.

The lines of dependency for food and other services are called food chains. Thus soil-oak-deer-Indian is a chain that has now been largely converted to soil-corn-cow-farmer. Each species, including ourselves, is a link in many chains. The deer eats a hundred plants other than oak, and the cow a hundred plants other than corn. Both, then, are links in a hundred chains. The pyramid is a tangle of chains so complex as to seem disorderly, yet the stability of the system proves it to be a highly organized structure. Its functioning depends on the co-operation and competition of its diverse parts.

In the beginning, the pyramid of life was low and squat; the food chains short and simple. Evolution has added layer after layer, link after link. Man is one of thousands of accretions to the height and complexity of the pyramid. Science has given us many doubts, but it has given us at least one certainty: the trend of evolution is to elaborate and diversify the biota.

Land, then, is not merely soil; it is a fountain of energy flowing through a circuit of soils, plants, and animals. Food chains are the living channels which conduct energy upward; death and decay return it to the soil. The circuit is not closed; some energy is dissipated in decay, some is added by absorption from the air, some is stored in soils, peats, and long-lived forests; but it is a sustained circuit, like a slowly augmented revolving fund of life. There is always a net loss by downhill wash, but this is normally small and offset by the decay of rocks. It is deposited in the ocean and, in the course of geological time, raised to form new lands and new pyramids.

The velocity and character of the upward flow of energy depend on the complex structure of the plant and animal community, much as the upward flow of sap in a tree depends on its complex cellular organization. Without this complexity, normal circulation would presumably not occur. Structure means the characteristic numbers, as well as the characteristic kinds and functions, of the component species. This interdependence between the complex structure of the land and its smooth functioning as an energy unit is one of its basic attributes.

When a change occurs in one part of the circuit, many other parts must adjust themselves to it. Change does not necessarily obstruct or divert the flow of energy; evolution is a long series of self-induced changes, the net result of which has been to elaborate the flow mechanism and to lengthen the circuit. Evolutionary changes, however, are usually slow and local. Man's invention of tools has enabled him to make changes of unprecedented violence, rapidity, and scope.

One change is in the composition of floras and faunas. The larger predators are lopped off the apex of the pyramid; food chains, for the first time in history, become shorter rather than longer. Domesticated species from other lands are substituted for wild ones, and wild ones are moved to new habitats. In this world-wide pooling of faunas and floras, some species get out of bounds as pests and diseases, others are extinguished. Such effects are seldom intended or foreseen; they represent unpredicted and often untraceable readjustments in the structure. Agricultural science is largely a race between the emergence of new pests and the emergence of new techniques for their control.

Another change touches the flow of energy through plants and animals and its return to the soil. Fertility is the ability of soil to receive, store, and release energy. Agriculture, by overdrafts on the soil, or by too radical a substitution of domestic for native species in the superstructure, may derange the channels of flow or deplete storage. Soils depleted of their storage, or of the organic matter which anchors it, wash away faster than they form. This is erosion.

Waters, like soil, are part of the energy circuit. Industry, by polluting waters or obstructing them with dams, may exclude the plants and animals necessary to keep energy in circulation.

Transportation brings about another basic change: the plants or animals grown in one region are now consumed and returned to the soil in another. Transportation taps the energy stored in rocks, and in the air, and uses it elsewhere; thus we fertilize the garden with nitrogen gleaned by the guano birds from the fishes of seas on the other side of the Equator. Thus the formerly localized and self-contained circuits are pooled on a world-wide scale.

The process of altering the pyramid for human occupation releases stored energy, and this often gives rise, during the pioneering period, to a deceptive exuberance of plant and animal life, both wild and tame. These releases of biotic capital tend to becloud or postpone the penalties of violence.

This thumbnail sketch of land as an energy circuit conveys three basic ideas:

(1) That land is not merely soil.
(2) That the native plants and animals kept the energy circuit open; others may or may not.
(3) That man-made changes are of a different order than evolutionary changes, and have effects more comprehensive than is intended or foreseen.

These ideas, collectively, raise two basic issues: Can the land adjust itself to the new order? Can the desired alterations be accomplished with less violence?

Biotas seem to differ in their capacity to sustain violent conversion. Western Europe, for example, carries a far different pyramid than Caesar found there. Some large animals are lost; swampy forests have become meadows or plowland; many new plants and animals are introduced, some of which escape as pests; the remaining natives are greatly changed in distribution and abundance. Yet the soil is still there and, with the help of imported nutrients, still fertile; the waters flow normally; the new structure seems to function and to persist. There is no visible stoppage or derangement of the circuit.

Western Europe, then, has a resistant biota. Its inner processes are tough, elastic, resistant to strain. No matter how violent the alterations, the pyramid, so far, has developed some new *modus vivendi* which preserves its habitability for man, and for most of the other natives.

Japan seems to present another instance of radical conversion without disorganization.

Most other civilized regions, and some as yet barely touched by civilization, display various stages of disorganization, varying from initial symptoms to advanced wastage. In Asia Minor and North Africa diagnosis is confused by climatic changes, which may have been either the cause or the effect of advanced wastage. In the United States the degree of disorganization varies locally; it is worst in the Southwest, the Ozarks, and parts of the South, and least in New England and the Northwest. Better land-uses may still arrest it in the less advanced regions. In parts of Mexico, South America, South Africa, and Australia a violent and accelerating wastage is in progress, but I cannot assess the prospects.

This almost world-wide display of disorganization in the land seems to be similar to disease in an animal, except that it never culminates in complete disorganization or death. The land recovers, but at some reduced level of complexity, and with a reduced carrying capacity for people, plants, and animals. Many biotas currently regarded as "lands of opportunity" are in fact already subsisting on exploitative agriculture, i.e. they have already exceeded their sustained carrying capacity. Most of South America is overpopulated in this sense.

In arid regions we attempt to offset the process of wastage by reclamation, but it is only too evident that the prospective longevity of reclamation projects is often short. In our own West, the best of them may not last a century.

The combined evidence of history and ecology seems to support one general deduction: the less violent the man-made changes, the greater the probability of successful readjustment in the pyramid. Violence, in turn, varies with human population density; a dense population requires a more violent conversion. In this respect, North America has a better chance for permanence than Europe, if she can contrive to limit her density.

This deduction runs counter to our current philosophy, which assumes that because a small increase in density enriched human life, that an indefinite increase will enrich it indefinitely. Ecology knows of no density relationship that holds for indefinitely wide limits. All gains from density are subject to a law of diminishing returns.

Whatever may be the equation for men and land, it is improbable that we as yet know all its terms. Recent discoveries in mineral and vitamin nutrition reveal unsuspected dependencies in the up-circuit: incredibly minute quantities of certain substances determine the value of soils to plants, of plants to animals. What of the down-circuit? What of the vanishing species, the preservation of which we now regard as an esthetic luxury? They helped build the soil; in what unsuspected ways may they be essential to its maintenance? Professor Weaver proposes that we use prairie flowers to reflocculate the wasting soils of the dust bowl; who knows for what purpose cranes and condors, otters and grizzlies may some day be used?

LAND HEALTH AND THE A-B CLEAVAGE

A land ethic, then, reflects the existence of an ecological conscience, and this in turn reflects a conviction of individual responsibility for the health of the land. Health is the capacity of the land for self-renewal. Conservation is our effort to understand and preserve this capacity.

Conservationists are notorious for their dissensions. Superficially these seem to add up to mere confusion, but a more careful scrutiny reveals a single plane of cleavage common to many specialized fields. In each field one group (A) regards the land as soil, and its function as commodity-production; another group (B) regards the land as a biota, and its function as something broader. How much broader is admittedly in a state of doubt and confusion.

In my own field, forestry, group A is quite content to grow trees like cabbages, with cellulose as the basic forest commodity. It feels no inhibition against violence; its ideology is agronomic. Group B, on the other hand, sees forestry as fundamentally different from agronomy because it employs natural species, and manages a natural environment rather than creating an artificial one. Group B prefers natural reproduction on principle. It worries on biotic as well as economic grounds about the loss of species like chestnut, and the threatened loss of the white pines. It worries about a whole series of secondary forest functions: wildlife, recreation, watersheds, wilderness areas. To my mind, Group B feels the stirrings of an ecological conscience.

In the wildlife field, a parallel cleavage exists. For Group A the basic commodities are sport and meat; the yardsticks of production are ciphers of take in pheasants and trout. Artificial propagation is acceptable as a permanent as well as a temporary recourse—if its unit costs permit. Group B, on the other hand, worries about a whole series of biotic side-issues. What is the cost in predators of producing a game crop? Should we have further recourse to exotics? How can management restore the shrinking species, like prairie grouse, already hopeless as shootable game? How can management restore the threatened rarities, like trumpeter swan and whooping crane? Can management principles be extended to wildflowers? Here again it is clear to me that we have the same A-B cleavage as in forestry.

In the larger field of agriculture I am less competent to speak, but there seem to be somewhat parallel cleavages. Scientific agriculture was actively developing before ecology was born, hence a slower penetration of ecological concepts might be expected. Moreover the farmer, by the very nature of his techniques, must modify the biota more radically than the forester or the wildlife manager. Nevertheless, there are many discontents in agriculture which seem to add up to a new vision of "biotic farming."

Perhaps the most important of these is the new evidence that poundage or tonnage is no measure of the food-value of farm crops; the products of fertile soil may be qualitatively as well as quantitatively superior. We can bolster poundage from depleted soils by pouring on imported fertility, but we are not necessarily bolstering food-value. The possible ultimate ramifications of this idea are so immense that I must leave their exposition to abler pens.

The discontent that labels itself "organic farming," while bearing some of the earmarks of a cult, is nevertheless biotic in its direction, particularly in its insistence on the importance of soil flora and fauna.

The ecological fundamentals of agriculture are just as poorly known to the public as in other fields of land-use. For example, few educated people realize that the marvelous advances in technique made during recent decades are improvements in the pump, rather than the well. Acre for acre, they have barely sufficed to offset the sinking level of fertility.

In all of these cleavages, we see repeated the same basic paradoxes: man the conqueror versus man the biotic citizen; science the sharpener of his sword versus science the searchlight on his universe; land the slave and servant versus land the collective organism. Robinson's injunction to Tristram may well be applied, at this juncture, to *Homo sapiens* as a species in geological time:

> Whether you will or not
> You are a King, Tristram, for you are one
> Of the time-tested few that leave the world,
> When they are gone, not the same place it was.
> Mark what you leave.

THE OUTLOOK

It is inconceivable to me that an ethical relation to land can exist without love, respect, and admiration for land, and a high regard for its value. By value, I of course mean something far broader than mere economic value; I mean value in the philosophical sense.

Perhaps the most serious obstacle impeding the evolution of a land ethic is the fact that our educational and economic system is headed away from, rather than toward, an intense consciousness of land. Your true modern is separated from the land by many middlemen, and by innumerable physical gadgets. He has no vital relation to it; to him it is the space between cities on which crops grow. Turn him loose for a day on the land, and if the spot does not happen to be a golf links or a "scenic" area, he is bored stiff. If crops could be raised by hydroponics instead of farming, it would suit him very well. Synthetic substitutes for wood, leather, wool, and other natural land products suit him better than the original. In short, land is something he has "outgrown."

Almost equally serious as an obstacle to a land ethic is the attitude of the farmer for whom the land is still an adversary, or a taskmaster that keeps him in slavery. Theoretically, the mechanization of farming ought to cut the farmer's chains, but whether it really does is debatable.

One of the requisites for an ecological comprehension of land is an understanding of ecology, and this is by no means co-extensive with "education"; in fact, much higher education seems deliberately to avoid ecological concepts. An understanding of ecology does not necessarily originate in courses bearing ecological labels; it is quite as likely to be labeled geography, botany, agronomy, history, or economics. This is as it should be, but whatever the label, ecological training is scarce.

The case for a land ethic would appear hopeless but for the minority which is in obvious revolt against these "modern" trends.

The "key-log" which must be moved to release the evolutionary process for an ethic is simply this: quit thinking about decent land-use as solely an economic problem. Examine each question in terms of what is ethically and esthetically right, as well as what is economically expedient. A thing is right when it tends to preserve the integrity, stability, and beauty of the biotic community. It is wrong when it tends otherwise.

It of course goes without saying that economic feasibility limits the tether of what can or cannot be done for land. It always has and it always will. The fallacy the economic determinists have tied around our collective neck, and which we now need to cast off, is the belief that economics determines all land-use. This is simply not true. An innumerable host of actions and attitudes, comprising perhaps the bulk of all land relations, is determined by the land-users' tastes and predilections, rather than by his purse. The bulk of all land relations hinges on investments of time, forethought, skill, and faith rather than on investments of cash. As a land-user thinketh, so is he.

I have purposely presented the land ethic as a product of social evolution because nothing so important as an ethic is ever "written." Only the most superficial student of history supposes that Moses "wrote" the Decalogue; it evolved in the minds of a thinking community, and Moses wrote a tentative summary of it for a "seminar." I say tentative because evolution never stops.

The evolution of a land ethic is an intellectual as well as emotional process. Conservation is paved with good intentions which prove to be futile, or even dangerous, because they are devoid of critical understanding either of the land, or of economic land-use. I think it is a truism that as the ethical frontier advances from the individual to the community, its intellectual content increases.

The mechanism of operation is the same for any ethic: social approbation for right actions: social disapproval for wrong actions.

By and large, our present problem is one of attitudes and implements. We are remodeling the Alhambra with a steam-shovel, and we are proud of our yardage. We shall hardly relinquish the shovel, which after all has many good points, but we are in need of gentler and more objective criteria for its successful use.

20

Rachel Carson (1907–1964)

From an early age Rachel Carson showed a talent for writing, publishing a number of award-winning pieces while still in her teens. While attending Pennsylvania College for Women (later Chatham College) Carson studied with a female biology professor who served as a role model for Carson in an era when few women sought careers in the sciences. Carson subsequently changed her major from English to Biology. And went on to earn a master's degree in marine biology from Johns Hopkins University in 1932. She continued to work toward her doctorate until her family's financial difficulties made it necessary for her to take a position with the U.S. Bureau of Fisheries (later incorporated into the U.S. Fish and Wildlife Service) in 1935. In 1941, Carson's first book, Under the Sea-Wind *was published. Although that book went largely unnoticed due to the Second World War, her next two books,* The Sea Around Us *(1951) and* The Edge of the Sea *(1955), were enormous critical and popular successes.*

In January 1958, Olga Owens Huckins wrote a letter to Carson about the devastating effect that the aerial spraying of DDT had recently had on a private bird sanctuary that she and her husband maintained in Duxbury, Massachusetts. As Carson wrote in the acknowledgments to Silent Spring, *that letter "brought my attention sharply back to a problem with which I had long been concerned. I then realized I must write this book." As Carson vividly shows in her book, the technology of the twentieth century often brings with it a myriad of unintended consequences that can have profound and widespread repercussions. The political impact of* Silent Spring *has been compared to such works as Thomas Paine's* Common Sense *and Harriet Beecher Stowe's* Uncle Tom's Cabin. *Carson's ability to make complex scientific information accessible to a general audience resulted in a public outcry for more research into the use of chemical pesticides, and led to a series of congressional hearings and the appointment of a Presidential Commission in 1963 to evaluate the dangers posed by chemical pesticides.*

"The Obligation to Endure"
From *Silent Spring*
(1962)

The history of life on earth has been a history of interaction between living things and their surroundings. To a large extent, the physical form and the habits of the earth's vegetation and its animal life have been molded by the environment. Considering the whole span of earthly time, the opposite effect, in which life actually modifies its surroundings, has been relatively slight. Only within the moment of time represented by the present century has one species—man—acquired significant power to alter the nature of his world.

During the past quarter century this power has not only increased to one of disturbing magnitude but it has changed in character. The most alarming of all man's assaults upon the environment is the contamination of air, earth, rivers, and sea with dangerous and even lethal materials. This pollution is for the most part irrecoverable; the chain of evil it initiates not only in the world that must support life but in living tissues is for the most part irreversible. In this now universal contamination of the environment, chemicals are the sinister and little-recognized partners of radiation in changing the very nature of the world—the very nature of its life. Strontium 90, released through nuclear explosions into the air, comes to earth in rain or drifts down as fallout, lodges in soil, enters into the grass or corn or wheat grown there, and in time takes up its abode in the bones of a human being, there to remain until his death. Similarly, chemicals sprayed on croplands or forests or gardens lie long in soil, entering into living organisms, passing from one to another in a chain of poisoning and death. Or they pass mysteriously by underground streams until they emerge and, through the alchemy of air and sunlight, combine into new forms that kill vegetation, sicken cattle, and work unknown harm on those who drink from once pure wells. As Albert Schweitzer has said, "Man can hardly even recognize the devils of his own creation."

It took hundreds of millions of years to produce the life that now inhabits the earth—eons of time in which that developing and evolving and diversifying life reached a state of adjustment and balance with its surroundings. The environment, rigorously shaping and directing the life it supported, contained elements that were hostile as well as supporting. Certain rocks gave out dangerous radiation; even within the light of the sun, from which all life draws its energy, there were short-wave radiations with power to injure. Given time—time not in years but in millennia—life adjusts, and a balance has been reached. For time is the essential ingredient; but in the modern world there is no time.

The rapidity of change and the speed with which new situations are created follow the impetuous and heedless pace of man rather than the

Reprinted by permission of Houghton Mifflin, "The Obligation to Endure," from *Silent Spring* by Rachel Carson. Copyright © 1962 by Rachel Carson, renewed 1990 by Roger Christie. All rights reserved.

deliberate pace of nature. Radiation is no longer merely the background radiation of rocks, the bombardment of cosmic rays, the ultraviolet of the sun that have existed before there was any life on earth; radiation is now the unnatural creation of man's tampering with the atom. The chemicals to which life is asked to make its adjustment are no longer merely the calcium and silica and copper and all the rest of the minerals washed out of the rocks and carried in rivers to the sea; they are the synthetic creations of man's inventive mind, brewed in his laboratories, and having no counterparts in nature.

To adjust to these chemicals would require time on the scale that is nature's; it would require not merely the years of a man's life but the life of generations. And even this, were it by some miracle possible, would be futile, for the new chemicals come from our laboratories in an endless stream; almost five hundred annually find their way into actual use in the United States alone. The figure is staggering and its implications are not easily grasped—500 new chemicals to which the bodies of men and animals are required somehow to adapt each year, chemicals totally outside the limits of biologic experience.

Among them are many that are used in man's war against nature. Since the mid-1940's over 200 basic chemicals have been created for use in killing insects, weeds, rodents, and other organisms described in the modern vernacular as "pests"; and they are sold under several thousand different brand names.

These sprays, dusts, and aerosols are now applied almost universally to farms, gardens, forests, and homes—nonselective chemicals that have the power to kill every insect, the "good" and the "bad," to still the song of birds and the leaping of fish in the streams, to coat the leaves with a deadly film, and to linger on in soil—all this though the intended target may be only a few weeds or insects. Can anyone believe it is possible to lay down such a barrage of poisons on the surface of the earth without making it unfit for all life? They should not be called "insecticides," but "biocides."

The whole process of spraying seems caught up in an endless spiral. Since DDT was released for civilian use, a process of escalation has been going on in which ever more toxic materials must be found. This has happened because insects, in a triumphant vindication of Darwin's principle of the survival of the fittest, have evolved super races immune to the particular insecticide used, hence a deadlier one has always to be developed—and then a deadlier one than that. It has happened also because, for reasons to be described later, destructive insects often undergo a "flareback," or resurgence, after spraying, in numbers greater than before. Thus the chemical war is never won, and all life is caught in its violent crossfire.

Along with the possibility of the extinction of mankind by nuclear war, the central problem of our age has therefore become the contamination of man's total environment with such substances of incredible potential for harm—substances that accumulate in the tissues of plants and animals and even penetrate the germ cells to shatter or alter the very material of heredity upon which the shape of the future depends.

Some would-be architects of our future look toward a time when it will be possible to alter the human germ plasm by design. But we may easily be doing so now by inadvertence, for many chemicals, like radiation, bring about gene mutations. It is ironic to think that man might determine his own future by something so seemingly trivial as the choice of an insect spray.

All this has been risked—for what? Future historians may well be amazed by our distorted sense of proportion. How could intelligent beings seek to control a few unwanted species by a method that contaminated the entire environment and brought the threat of disease and death even to their own kind? Yet this is precisely what we have done. We have done it, moreover, for reasons that collapse the moment we examine them. We are told that the enormous and expanding use of pesticides is necessary to maintain farm production. Yet is our real problem not one of *overproduction*? Our farms, despite measures to remove acreages from production and to pay farmers *not* to produce, have yielded such a staggering excess of crops that the American taxpayer in 1962 is paying out more than one billion dollars a year as the total carrying cost of the surplus-food storage program. And is the situation helped when one branch of the Agriculture Department tries to reduce production while another states, as it did in 1958, "It is believed generally that reduction of crop acreages under provisions of the Soil Bank will stimulate interest in use of chemicals to obtain maximum production on the land retained in crops."

All this is not to say there is no insect problem and no need of control. I am saying, rather, that control must be geared to realities, not to mythical situations, and that the methods employed must be such that they do not destroy us along with the insects.

The problem whose attempted solution has brought such a train of disaster in its wake is an accompaniment of our modern way of life. Long before the age of man, insects inhabited the earth—a group of extraordinarily varied and adaptable beings. Over the course of time since man's advent, a small percentage of the more than half a million species of insects have come into conflict with human welfare in two principal ways: as competitors for the food supply and as carriers of human disease.

Disease-carrying insects become important where human beings are crowded together, especially under conditions where sanitation is poor, as in time of natural disaster or war or in situations of extreme poverty and deprivation. Then control of some sort becomes necessary. It is a sobering fact, however, as we shall presently see, that the method of massive chemical control has had only limited success, and also threatens to worsen the very conditions it is intended to curb.

Under primitive agricultural conditions the farmer had few insect problems. These arose with the intensification of agriculture—the devotion of immense acreages to a single crop. Such a system set the stage for explosive increases in specific insect populations. Single-crop farming does not take advantage of the principles by which nature works; it is agriculture as an engineer might conceive it to be. Nature has introduced great variety into the landscape, but man has displayed a passion for simplifying it. Thus he undoes the built-in checks and balances by which nature holds the species within bounds. One important natural check is a limit on the amount of suitable habitat for each species. Obviously then, an insect that lives on wheat can build up its population to much higher levels on a farm devoted to wheat than on one in which wheat is intermingled with other crops to which the insect is not adapted.

The same thing happens in other situations. A generation or more ago, the towns of large areas of the United States lined their streets with the noble elm tree. Now the beauty they hopefully created is threatened with complete destruction as disease sweeps through the elms, carried by a beetle that would have only limited chance to build up large populations and to spread from tree to tree if the elms were only occasional trees in a richly diversified planting.

Another factor in the modern insect problem is one that must be viewed against a background of geologic and human history: the spreading of thousands of different kinds of organisms from their native homes to invade new territories. This worldwide migration has been studied and graphically described by the British ecologist Charles Elton in his recent book *The Ecology of Invasions.* During the Cretaceous Period, some hundred million years ago, flooding seas cut many land bridges between continents and living things found themselves confined in what Elton calls "colossal separate nature reserves." There, isolated from others of their kind, they developed many new species. When some of the land masses were joined again, about 15 million years ago, these species began to move out into new territories—a movement that is not only still in progress but is now receiving considerable assistance from man.

The importation of plants is the primary agent in the modern spread of species, for animals have almost invariably gone along with the plants, quarantine being a comparatively recent and not completely effective innovation. The United States Office of Plant Introduction alone has introduced almost 200,000 species and varieties of plants from all over the world. Nearly half of the 180 or so major insect enemies of plants in the United States are accidental imports from abroad, and most of them have come as hitchhikers on plants.

In new territory, out of reach of the restraining hand of the natural enemies that kept down its numbers in its native land, an invading plant or animal is able to become enormously abundant. Thus it is no accident that our most troublesome insects are introduced species.

These invasions, both the naturally occurring and those dependent on human assistance, are likely to continue indefinitely. Quarantine and massive chemical campaigns are only extremely expensive ways of buying time. We are faced, according to Dr. Elton, "with a life-and-death need not just to find new technological means of suppressing this plant or that animal"; instead we need the basic knowledge of animal populations and their relations to their surroundings that will "promote an even balance and damp down the explosive power of outbreaks and new invasions."

Much of the necessary knowledge is now available but we do not use it. We train ecologists in our universities and even employ them in our governmental agencies but we seldom take their advice. We allow the chemical death rain to fall as though there were no alternative, whereas in fact there are many, and our ingenuity could soon discover many more if given opportunity.

Have we fallen into a mesmerized state that makes us accept as inevitable that which is inferior or detrimental, as though having lost the will or the vision to demand that which is good? Such thinking, in the words of the ecologist Paul Shepard, "idealizes life with only its head out of water, inches above the limits of toleration of the corruption of its own environment Why should we tolerate a diet of weak poisons, a home in insipid surroundings, a circle of acquaintances who are not quite our enemies, the noise of motors with just enough relief to prevent insanity? Who would want to live in a world which is just not quite fatal?"

Yet such a world is pressed upon us. The crusade to create a chemically sterile, insect-free world seems to have engendered a fanatic zeal on the part of many specialists and most of the so-called control agencies. On every hand there is evidence that those engaged in spraying operations exercise a ruthless power. "The regulatory entomologists . . . function as prosecutor, judge and jury, tax assessor and collector and sheriff to enforce their own orders," said Connecticut entomologist Neely Turner. The most flagrant abuses go unchecked in both state and federal agencies.

It is not my contention that chemical insecticides must never be used. I do contend that we have put poisonous and biologically potent chemicals indiscriminately into the hands of persons largely or wholly ignorant of their potentials for harm. We have subjected enormous numbers of people to contact with these poisons, without their consent and often without their knowledge. If the Bill of Rights contains no guarantee that a citizen shall be secure against lethal poisons distributed either by private individuals or by public officials, it is surely only because our forefathers, despite their considerable wisdom and foresight, could conceive of no such problem.

I contend, furthermore, that we have allowed these chemicals to be used with little or no advance investigation of their effect on soil, water, wildlife, and man himself. Future generations are unlikely to condone our lack of prudent concern for the integrity of the natural world that supports all life.

There is still very limited awareness of the nature of the threat. This is an era of specialists, each of whom sees his own problem and is unaware of or intolerant of the larger frame into which it fits. It is also an era dominated by industry, in which the right to make a dollar at whatever cost is seldom challenged. When the public protests, confronted with some obvious evidence of damaging results of pesticide applications, it is fed little tranquilizing pills of half truth. We urgently need an end to these false assurances, to the sugar coating of unpalatable facts. It is the public that is being asked to assume the risks that the insect controllers calculate. The public must decide whether it wishes to continue on the present road, and it can do so only when in full possession of the facts. In the words of Jean Rostand, "The obligation to endure gives us the right to know."

21

The Wilderness Act (1964)

While writers such as John Muir and Bob Marshall had long argued that there was a need to protect wilderness areas, it was not until the passage of the Wilderness Act in 1964 that federal policy makers took a significant step beyond the "wise use" philosophy of the Progressive era. In 1956 the executive director of the Wilderness Society, Howard Zahniser, proposed legislation that would permanently protect millions of acres wilderness areas located in the national parks and forest reserves. Senator Hubert H. Humphrey (D-MN) and Representative John Saylor (R-PA) introduced such a bill the following year. After numerous revisions, the bill was signed into law by President Lyndon B. Johnson on September 3, 1964. While the law's final definition of wilderness was more limited than many wilderness advocates had hoped, and the bill itself left the thorny issue of designating future wilderness areas largely unresolved, it was still a legislative milestone in wilderness protection and codified the desirability of creating a National Wilderness Preservation System.

The Wilderness Act of 1964
Public Law 88–577 (16 U.S.C. 1131–1136)
88th Congress, Second Session
September 3,1964

An Act to establish a National Wilderness Preservation System for *the permanent* good *of the whole people, and for other purposes.*

Be it enacted by the Senate and House of Representatives of the United States of America in Congress assembled,

Short Title

Section 1. This Act may be cited as the "Wilderness Act."

WILDERNESS SYSTEM ESTABLISHED—STATEMENT OF POLICY

Sec. 2. (a) In order to assure that an increasing population, accompanied by expanding settlement and growing mechanization, does not occupy and modify all areas within the United States and its possessions, leaving no lands designated for preservation and protection in their natural condition, it is hereby declared to be the policy of the Congress to secure for the American people of present and future generations the benefits of an enduring resource of wilderness. For this purpose there is hereby established a National Wilderness Preservation System to be composed of federally owned areas designated by Congress as "wilderness areas," and these shall be administered for the use and enjoyment of the American people in such manner as will leave them unimpaired for future use and enjoyment as wilderness; and no Federal lands shall be designated as "wilderness areas" except as provided for in this Act or by a subsequent Act.

(b) The inclusion of an area in the National Wilderness Preservation System notwithstanding, the area shall continue to be managed by the Department and agency having jurisdiction thereover immediately before its inclusion in the National Wilderness Preservation System unless otherwise provided by Act of Congress. No appropriation shall be available for the payment of expenses or salaries for the administration of the National Wilderness Preservation System as a separate unit nor shall any appropriations be available for additional personnel stated as being required solely for the purpose of managing or administering areas solely because they are included within the National Wilderness Preservation System.

DEFINITION OF WILDERNESS

(c) A wilderness, in contrast with those areas where man and his own works dominate the landscape, is hereby recognized as an area where the earth and its community of life are untrammeled by man, where man himself is a visitor who does not remain. An area of wilderness is further defined to mean in this Act an area of undeveloped Federal land retaining its primeval character and influence, without permanent improvements or human habitation, which is protected and managed so as to preserve its natural conditions and which (1) generally appears to have been affected primarily by the forces of nature, with the imprint of mans work substantially unnoticeable; (2) has outstanding opportunities for solitude or a primitive and unconfined type of recreation; (3) has at least five thousand acres of land or is of sufficient size as to make practicable its preservation and use in an unimpaired condition; and (4) may also contain ecological, geological, or other features of scientific, educational, scenic, or historical value.

22

Lynn White, Jr. (1907–1987)

Lynn White, Jr. was a medieval scholar and former college president (Mills College, 1943–1958) who was instrumental in establishing the Center for Medieval and Renaissance Studies at UCLA in 1962. White's primary interest was the historical development of technology and the effect that technology had on the societies of Europe during the medieval period. He is probably best known, however, for "The Historical Roots of Our Ecologic Crisis," which appeared in the magazine Science *in 1967. In this short article White argued that humankind's destructive impact on the environment was largely attributable to the link of Western science and technology to a Judeo-Christian tradition that emphasized "man's transcendence of, and rightful mastery over, nature." The result, said White, was a modern ecological crisis for which "Christianity bears a huge measure of guilt." According to White, the remedy for this was not a renunciation of Christianity, but the adoption of an alternative Christian view toward nature, that of Francis of Assisi. White's article is still capable of eliciting strong reactions (both positive and negative), but it also helped to initiate a significant reappraisal of the role that religion should play in the formation of environmental ethics.*

"The Historical Roots of Our Ecologic Crisis" (1967)

A conversation with Aldous Huxley not infrequently put one at the receiving end of an unforgettable monologue. About a year before his lamented death he was discoursing on a favorite topic: Man's unnatural treatment of nature and its sad results. To illustrate his point he told how, during the previous summer, he had returned to a little valley in England where he had spent many happy months as a child. Once it had been composed of delightful grassy glades; now it was becoming overgrown with unsightly brush because

Reprinted with permission from *Science*, Vol. 155, No. 3767 (March 10, 1967). Copyright © 1967, American Association for the Advancement of Science (AAAS).

the rabbits that formerly kept such growth under control had largely succumbed to a disease, myxomatosis, that was deliberately introduced by the local farmers to reduce the rabbits' destruction of crops. Being something of a Philistine, I could be silent no longer, even in the interests of great rhetoric. I interrupted to point out that the rabbit itself had been brought as a domestic animal to England in 1176, presumably to improve the protein diet of the peasantry.

All forms of life modify their contexts. The most spectacular and benign instance is doubtless the coral polyp. By serving its own ends, it has created a vast undersea world favorable to thousands of other kinds of animals and plants. Ever since man became a numerous species he has affected his environment notably. The hypothesis that his fire-drive method of hunting created the world's great grasslands and helped to exterminate the monster mammals of the Pleistocene from much of the globe is plausible, if not proved. For 6 millennia at least, the banks of the lower Nile have been a human artifact rather than the swampy African jungle which nature, apart from man, would have made it. The Aswan Dam, flooding 5000 square miles, is only the latest stage in a long process. In many regions terracing or irrigation, overgrazing, the cutting of forests by Romans to build ships to fight Carthaginians or by Crusaders to solve the logistics problems of their expeditions, have profoundly changed some ecologies. Observation that the French landscape fails into two basic types, the open fields of the north and the *bocage* of the south and west, inspired Marc Bloch to undertake his classic study of medieval agricultural methods. Quite unintentionally, changes in human ways often affect nonhuman nature. It has been noted, for example, that the advent of the automobile eliminated huge flocks of sparrows that once fed on the horse manure littering every street.

The history of ecologic change is still so rudimentary that we know little about what really happened, or what the results were. The extinction of the European aurochs as late as 1627 would seem to have been a simple case of overenthusiastic hunting. On more intricate matters it often is impossible to find solid information. For a thousand years or more the Frisians and Hollanders have been pushing back the North Sea, and the process is culminating in our own time in the reclamation of the Zuider Zee. What, if any, species of animals, birds, fish, shore life, or plants have died out in the process? In their epic combat with Neptune have the Netherlanders overlooked ecological values in such a way that the quality of human life in the Netherlands has suffered? I cannot discover that the questions have ever been asked, much less answered.

People, then, have often been a dynamic element in their own environment, but in the present state of historical scholarship we usually do not know exactly when, where, or with what effects man-induced changes came. As we enter the last third of the 20th century, however, concern for the problem of ecologic backlash is mounting feverishly. Natural science, conceived as the effort to understand the nature of things, had flourished in several eras and among several peoples. Similarly there had been an age-old accumulation of technological skills, sometimes growing rapidly, sometimes slowly. But it was not until about four generations ago that Western Europe and North America arranged a marriage between science and technology, a union of the theoretical and the empirical approaches to our natural environment. The emergence in widespread practice of the Baconian creed that scientific knowledge means technological power over nature can scarcely be dated before about 1850, save in the chemical industries, where it is anticipated in the 18th century. Its acceptance as a normal pattern of action may mark the greatest event in human history since the invention of agriculture, and perhaps in nonhuman terrestrial history as well.

Almost at once the new situation forced the crystallization of the novel concept of ecology; indeed, the word ecology first appeared in the English language in 1873. Today, less than a century later, the impact of our race upon the environment has so increased in force that it has changed in essence. When the first cannons were fired, in the early 14th century, they affected ecology by sending workers scrambling to the forests and mountains for more potash, sulfur, iron ore, and charcoal, with some resulting erosion and deforestation. Hydrogen bombs are of a different order: a war fought with them might alter the genetics of all life on this planet. By 1285 London had a smog problem arising from the burning of soft coal, but our present combustion of fossil fuels threatens to change the chemistry of the globe's atmosphere as a whole, with consequences which we are only beginning to guess. With the population explosion, the carcinoma of planless urbanism, the now geological deposits of sewage and garbage, surely no creature other than man has ever managed to foul its nest in such short order.

There are many calls to action, but specific proposals, however worthy as individual items, seem too partial, palliative, negative: ban the bomb, tear down the billboards, give the Hindus contraceptives and tell them to eat their sacred cows. The simplest solution to any suspect change is, of course, to stop it, or, better yet, to revert to a romanticized past: make those ugly gasoline stations look like Anne Hathaway's cottage or (in the Far West) like ghost-town saloons. The "wilderness area" mentality invariably advocates deep-freezing an ecology, whether San Gimignano or the High Sierra, as it was before the first Kleenex was dropped. But neither atavism nor prettification will cope with the ecologic crisis of our time.

What shall we do? No one yet knows. Unless we think about fundamentals, our specific measures may produce new backlashes more serious than those they are designed to remedy.

As a beginning we should try to clarify our thinking by looking, in sonic historical depth, at the presuppositions that underlie modern technology and science. Science was traditionally aristocratic, speculative, intellectual in intent: technology was lower-class, empirical, action-oriented. The quite sudden fusion of these two, towards the middle of the 19th century, is surely related to the slightly prior and contemporary democratic revolutions which, by reducing social barriers, tended to assert a functional unity of brain and hand. Our ecologic crisis is the product of an emerging, entirely novel, democratic culture. The issue is whether a democratized world can survive its own implications. Presumably we cannot unless we rethink our axioms.

The Western Traditions of Technology and Science

One thing is so certain that it seems stupid to verbalize it: both modern technology and modern science are distinctively *Occidental*. Our technology has absorbed elements from all over the world, notably from China; yet everywhere today, whether in Japan or in Nigeria, successful technology is Western. Our science is the heir to all the sciences of the past, especially perhaps to the work of the great Islamic scientists of the Middle Ages, who so often outdid the ancient Greeks in skill and perspicacity: al-Rāzī in medicine, for example; or ibn-al-Haytham in optics; or Omar Khayyām in mathematics. Indeed, not a few works of such geniuses seem to have vanished in the original Arabic and to survive only in medieval Latin translations that helped to lay the foundations for later Western developments. Today, around the globe, all significant science is Western in style and method, whatever the pigmentation or language of the scientists.

A second pair of facts is less well recognized because they result from quite recent historical scholarship. The leadership of the West, both in technology and in science, is far older than the so-called Scientific Revolution of the 17th century or the so-called Industrial Revolution of the 18th century. These terms are in fact outmoded and obscure the true nature of what they try to describe—significant stages in two long and separate developments. By A.D. 1000 at the latest—and perhaps, feebly, as much as 200 years earlier—the West began to apply water power to industrial processes other than milling grain. This was followed in the late 12th century by the harnessing of wind power. From simple beginnings, but with remarkable consistency of style, the West rapidly expanded its skills in the development of power machinery, labor-saving devices, and automation. Those who doubt should contemplate that most monumental achievement in the history of automation: the weight-driven mechanical clock, which appeared in two forms in the early 14th century. Not in craftsmanship but in basic technological capacity, the Latin West of the later Middle Ages far outstripped its elaborate, sophisticated, and esthetically magnificent sister cultures, Byzantium and Islam. In 1444 a great Greek ecclesiastic, Bessarion, who had gone to Italy, wrote a letter to a prince in Greece. He is amazed by the superiority of Western ships, arms, textiles, glass. But above all he is astonished by the spectacle of waterwheels sawing timbers and pumping the bellows of blast furnaces. Clearly, he had seen nothing of the sort in the Near East.

By the end of the 15th century the technological superiority of Europe was such that its small, mutually hostile nations could spill out over all the rest of the world, conquering, looting, and colonizing. The symbol of this technological superiority is the fact that Portugal, one of the weakest states of the Occident was able to become, and to remain for a century, mistress of the

East Indies. And we must remember that the technology of Vasco da Gama and Albuquerque was built by pure empiricism, drawing remarkably little support or inspiration from science.

In the present-day vernacular understanding, modern science is supposed to have begun in 1543, when both Copernicus and Vesalius published their great works. It is no derogation of their accomplishments, however, to point out that such structures as the *Fabrica* and the *De revolutionibus* do not appear overnight. The distinctive Western tradition of science, in fact, began in the late 11th century with a massive movement of translation of Arabic and Greek scientific works into Latin. A few notable books—Theophrastus, for example—escaped the West's avid new appetite for science, but within less than 200 years effectively the entire corpus of Greek and Muslim science was available in Latin, and was being eagerly read and criticized in the new European universities. Out of criticism arose new observation, speculation, and increasing distrust of ancient authorities. By the late 13th century Europe had seized global scientific leadership from the faltering hands of Islam. It would be as absurd to deny the profound originality of Newton, Galileo, or Copernicus as to deny that of the 14th century scholastic scientists like Buridan or Oresme on whose work they built. Before the 11th century, science scarcely existed in the Latin West, even in Roman times. From the 11th century onward, the scientific sector of Occidental culture has increased in a steady crescendo.

Since both our technological and our scientific movements got their start, acquired their character, and achieved world dominance in the Middle Ages, it would seem that we cannot understand their nature or their present impact upon ecology without examining fundamental medieval assumptions and developments.

MEDIEVAL VIEW OF MAN AND NATURE

Until recently, agriculture has been the chief occupation even in "advanced" societies; hence, any change in methods of tillage has much importance. Early plows, drawn by two oxen, did not normally turn the sod but merely scratched it. Thus, cross-plowing was needed and fields tended to be squarish. In the fairly light soils and semiarid climates of the Near East and Mediterranean, this worked well. But such a plow was inappropriate to the wet climate and often sticky soils of northern Europe. By the latter part of the 7th century after Christ, however, following obscure beginnings, certain northern peasants were using an entirely new kind of plow, equipped with a vertical knife to cut the line of the furrow, a horizontal share to slice under the sod, and a moldboard to turn it over. The friction of this plow with the soil was so great that it normally required not two but eight oxen. It attacked the land with such violence that cross-plowing was not needed, and fields tended to be shaped in long strips.

In the days of the scratch-plow, fields were distributed generally in units capable of supporting a single family. Subsistence fanning was the presupposition. But no peasant owned eight oxen: to use the new and more efficient plow, peasants pooled their oxen to form large plow-teams, originally receiving (it would appear) plowed strips in proportion to their contribution. Thus, distribution of land was based no longer on the needs of a family but, rather, on the capacity of a power machine to till the earth. Man's relation to the soil was profoundly changed. Formerly man had been part of nature; now he was the exploiter of nature. Nowhere else in the world did farmers develop any analogous agricultural implement. Is it coincidence that modern technology, with its ruthlessness toward nature, has so largely been produced by descendants of these peasants of northern Europe?

This same exploitive attitude appears slightly before A.D. 830 in Western illustrated calendars. In older calendars the months were shown as passive personifications. The new Frankish calendars, which set the style for the Middle Ages, are very different: they show men coercing the world around them—plowing, harvesting, chopping trees, butchering pigs. Man and nature are two things, and man is master.

These novelties seem to be in harmony with larger intellectual patterns. What people do about their ecology depends on what they think about themselves in relation to things around them. Human ecology is deeply conditioned by beliefs about our nature and destiny—that is, by religion. To Western eyes this is very evident in, say, India or Ceylon. It is equally true of ourselves and of our medieval ancestors.

The victory of Christianity over paganism was the greatest psychic revolution in the history of our culture. It has become fashionable today to say that, for better or worse, we live in "the post-Christian age." Certainly the forms of our thinking and language have largely ceased to be Christian, but to my eye the substance often remains amazingly akin to that of the past. Our daily habits of action, for example, are dominated by an implicit faith in perpetual progress which was unknown either to Greco-Roman antiquity or to the Orient. It is rooted in, and is indefensible apart from, Judeo-Christian teleology. The fact that Communists share it merely helps to show what can be demonstrated on many other grounds: that Marxism, like Islam, is a Judeo-Christian heresy. We continue today to live, as we have lived for about 1700 years, very largely in a context of Christian axioms.

What did Christianity tell people about their relations with the environment? While many of the world's mythologies provide stories of creation, Greco-Roman mythology was singularly incoherent in this respect. Like Aristotle, the intellectuals of the ancient West denied that the visible world had had a beginning. Indeed, the idea of a beginning was impossible in the framework of their cyclical notion of time. In sharp contrast, Christianity inherited from Judaism not only a concept of time as nonrepetitive and linear but also a striking story of creation. By gradual stages a loving and all-powerful God had created light and darkness, the heavenly bodies, the earth and all its plants, animals, birds, and fishes. Finally, God had created Adam and, as an afterthought, Eve to keep man from being lonely. Man named all the animals, thus establishing his dominance over them. God planned all of this explicitly for man's benefit and rule: no item in the physical creation had any purpose save to serve man's purposes. And, although man's body is made of clay, he is not simply part of nature: he is made in God's image.

Especially in its Western form, Christianity is the most anthropocentric religion the world has seen. As early as the 2nd century both Tertulian and Saint Irenaeus of Lyons were insisting that when God shaped Adam he was foreshadowing the image of the incarnate Christ, the Second Adam. Man shares, in great measure, God's transcendence of nature. Christianity, in absolute contrast to ancient paganism and Asia's religions (except, perhaps, Zoroastrianism), not only established a dualism of man and nature but also insisted that it is God's will that man exploit nature for his proper ends.

At the level of the common people this worked out in an interesting way. In Antiquity every tree, every spring, every stream, every bill had its own genius loci, its guardian spirit. These spirits were accessible to men, but were very unlike men; centaurs, fauns, and mermaids show their ambivalence. Before one cut a tree, mined a mountain, or dammed a brook, it was important to placate the spirit in charge of that particular situation, and to keep it placated. By destroying pagan animism, Christianity made it possible to exploit nature in a mood of indifference to the feelings of natural objects.

It is often said that for animism the Church substituted the cult of saints. True; but the cult of saints is functionally quite different from animism. The saint is not in natural objects; he may have special shrines, but his citizenship is in heaven. Moreover, a saint is entirely a man; he can be approached in human terms. In addition to saints, Christianity of course also bad angels and demons inherited from Judaism and perhaps, at one remove, from Zoroastrianism. But these were all as mobile as the saints themselves. The spirits in natural objects, which formerly had protected nature from man, evaporated. Man's effective monopoly on spirit in this world was confirmed, and the old inhibitions to the exploitation of nature crumbled.

When one speaks in such sweeping terms, a note of caution is in order. Christianity is a complex faith, and its consequences differ in differing contexts. What I have said may well apply to the medieval West, where in fact technology made spectacular advances. But the Greek East, a highly civilized realm of equal Christian devotion, seems to have produced no marked technological innovation after the late 7th century, when Greek fire was invented. The key to the contrast may perhaps be found in a difference in the tonality of piety and thought which students of comparative theology find between the Greek and the Latin Churches. The Greeks believed that sin was intellectual blindness, and that salvation was found in illumination, orthodoxy—that is, clear thinking. The Latins, on the other hand, felt that sin was moral evil, and that salvation was to be found in right conduct. Eastern theology has been intellectualist. Western theology has been voluntarist. The Greek saint contemplates; the Western saint acts. The implications of Christianity for the conquest of nature would emerge more easily in the Western atmosphere.

The Christian dogma of creation, which is found in the first clause of all the Creeds, has another meaning for our comprehension of today's ecologic crisis.

By revelation, God had given man the Bible, the Book of Scripture. But since God had made nature, nature also must reveal the divine mentality. The religious study of nature for the better understanding of God was known as natural theology. In the early Church, and always in the Greek East, nature was conceived primarily as a symbolic system through which God speaks to men: the ant is a sermon to sluggards; rising flames are the symbol of the soul's aspiration. This view of nature was essentially artistic rather than scientific. While Byzantium preserved and copied great numbers of ancient Greek scientific texts, science as we conceive it could scarcely flourish in such an ambience.

However, in the Latin West by the early 13th century natural theology was following a very different bent. It was ceasing to be the decoding of the physical symbols of God's communication with man and was becoming the effort to understand God's mind by discovering how his creation operates. The rainbow was no longer simply a symbol of hope first sent to Noah after the Deluge: Robert Grosseteste, Friar Roger Bacon, and Theodoric of Freiberg produced startlingly sophisticated work on the optics of the rainbow, but they did it as a venture in religious understanding. From the 13th century onward, up to and including Leibnitz and Newton, every major scientist, in effect, explained his motivations in religious terms. Indeed, if Galileo had not been so expert an amateur theologian he would have got into far less trouble: the professionals resented his intrusion. And Newton seems to have regarded himself more as a theologian than as a scientist. It was not until the late 18th century that the hypothesis of God became unnecessary to many scientists.

It is often hard for the historian to judge, when men explain why they are doing what they want to do, whether they are offering real reasons or merely culturally acceptable reasons. The consistency with which scientists during the long formative centuries of Western science said that the task and the reward of the scientist was "to think God's thoughts after him" leads one to believe that this was their real motivation. If so, then modern Western science was cast in a matrix of Christian theology. The dynamism of religious devotion, shaped by the Judeo-Christian dogma of creation, gave it impetus.

AN ALTERNATIVE CHRISTIAN VIEW

We would seem to be headed toward conclusions unpalatable to many Christians. Since both science and technology are blessed words in our contemporary vocabulary, some may be happy at the notions, first, that, viewed historically, modern science is an extrapolation of natural theology and, second, that modern technology is at least partly to be explained as an Occidental, voluntarist realization of the Christian dogma of man's transcendence of, and rightful mastery over, nature. But, as we now recognize, somewhat over a century ago science and technology—hitherto quite separate activities—joined to give mankind powers which, to judge by many of the ecologic effects, are out of control. If so, Christianity bears a huge burden of guilt.

I personally doubt that disastrous ecologic backlash can be avoided simply by applying to our problems more science and more technology. Our science and technology have grown out of Christian attitudes toward man's relation to nature which are almost universally held not only by Christians and neo-Christians but also by those who fondly regard themselves as post-Christians. Despite Copernicus, all the cosmos rotates around our little globe. Despite Darwin, we are not, in our hearts, part of the natural process. We are superior to nature, contemptuous of it, willing to use it for our slightest whim. The newly elected Governor of California, like myself a churchman but less troubled than I, spoke for the Christian tradition when he said (as is alleged), "when you've seen one redwood tree, you've seen them all." To a Christian a tree can be no more than a physical fact. The whole concept of the sacred grove is alien to Christianity and to the ethos of the West. For nearly 2 millennia Christian missionaries have been chopping down sacred groves, which are idolatrous because they assume spirit in nature.

What we do about ecology depends on our ideas of the man-nature relationship. More science and more technology are not going to get us out of the present ecologic crisis until we find a new religion, or rethink our old one. The beatniks, who are the basic revolutionaries of our time, show a sound instinct in their affinity for Zen Buddhism, which conceives of the man-nature relationship as very nearly the mirror image of the Christian view. Zen, however, is as deeply conditioned by Asian history as Christianity is by the experience of the West, and I am dubious of its viability among us.

Possibly we should ponder the greatest radical in Christian history since Christ: Saint Francis of Assisi. The prime miracle of Saint Francis is the fact that he did not end at the stake, as many of his left-wing followers did. He was so clearly heretical that a General of the Franciscan Order, Saint Bonaventura, a great and perceptive Christian, tried to suppress the early accounts of Franciscanism. The key to an understanding of Francis is his belief in the virtue of humility—not merely for the individual but for man as a species. Francis tried to depose man from his monarchy over creation and set up a democracy of all God's creatures. With him the ant is no longer simply a homily for the lazy, flames a sign of the thrust of the soul toward union with God; now they are Brother Ant and Sister Fire, praising the Creator in their own ways as Brother Man does in his.

Later commentators have said that Francis preached to the birds as a rebuke to men who would not listen. The records do not read so: he urged the little birds to praise God, and in spiritual ecstasy they flapped their wings and chirped rejoicing. Legends of saints, especially the Irish saints, had long told of their dealings with animals but always, I believe, to show their human dominance over creatures. With Francis it is different. The land around Gubbio in the Apennines was being ravaged by a fierce wolf. Saint Francis, says the legend, talked to the wolf and persuaded him of the error of his ways. The wolf repented, died in the odor of sanctity, and was buried in consecrated ground.

What Sir Steven Ruciman calls "the Franciscan doctrine of the animal soul" was quickly stamped out. Quite possibly it was in part inspired, consciously or unconsciously, by the belief in reincarnation held by the

Cathar heretics who at that time teemed in Italy and southern France, and who presumably had got it originally from India. It is significant that at just the same moment, about 1200, traces of metempsychosis are found also in western Judaism, in the Provençal *Cabbala*. But Francis held neither to transmigration of souls nor to pantheism. His view of nature and of man rested on a unique sort of pan-psychism of all things animate and inanimate, designed for the glorification of their transcendent Creator, who, in the ultimate gesture of cosmic humility, assumed flesh, lay helpless in a manger, and hung dying on a scaffold.

I am not suggesting that many contemporary Americans who are concerned about our ecologic crisis will be either able or willing to counsel with wolves or exhort birds. However, the present increasing disruption of the global environment is the product of a dynamic technology and science which were originating in the Western medieval world against which Saint Francis was rebelling in so original a way. Their growth cannot be understood historically apart from distinctive attitudes toward nature which are deeply grounded in Christian dogma. The fact that most people do not think of these attitudes as Christian is irrelevant. No new set of basic values has been accepted in our society to displace those of Christianity. Hence we shall continue to have a worsening ecologic crisis until we reject the Christian axiom that nature has no reason for existence save to serve man.

The greatest spiritual revolutionary in Western history, Saint Francis, proposed what he thought was an alternative Christian view of nature and man's relation to it: he tried to substitute the idea of the equality of all creatures, including man, for the idea of man's limitless rule of creation. He failed. Both our present science and our present technology are so tinctured with orthodox Christian arrogance toward nature that no solution for our ecologic crisis can be expected from them alone. Since the roots of our trouble are so largely religious, the remedy must also be essentially religious, whether we call it that or not. We must rethink and refeel our nature and destiny. The profoundly religious, but heretical, sense of the primitive Franciscans for the spiritual autonomy of all parts of nature may point a direction. I propose Francis as a patron saint for ecologists.

23

GARRETT HARDIN (1915–2003)

After earning his doctorate in microbiology at Stanford University in 1941, Hardin went on to work as a plant biologist at the Carnegie Institution of Washington. In 1946 he took a faculty position at the University of California, Santa Barbara, where he taught in the Department of Human Ecology for over thirty years. Much of Hardin's work was in the field of bioethics, and he played a prominent role in the public debate over issues such as abortion, population control, and immigration reform. The publication of Hardin's best known essay, "The Tragedy of the Commons," focused attention on the ecological and ethical dilemmas of population control. His proposed solution to overpopulation, "mutual coercion, mutually agreed upon by the majority of the people affected," while still controversial, continues to be a touchstone of any debate over the allocation of commonly held resources. Hardin continued to stress the dangers of overpopulation in a number of later articles and books including Avoiding the Tragedy of the Commons *(1995) and* The Ostrich Factor: Our Population Myopia *(1999).*

"The Tragedy of the Commons" (1968)

At the end of a thoughtful article on the future of nuclear war, Wiesner and York (1) concluded that: "Both sides in the arms race are . . . confronted by the dilemma of steadily increasing military power and steadily decreasing national security. *It is our considered professional judgment that this dilemma has no technical solution.* If the great powers continue to look for solutions in the area of science and technology only, the result will be to worsen the situation."

I would like to focus your attention not on the subject of the article (national security in a nuclear world) but on the kind of conclusion they reached, namely that there is no technical solution to the problem. An

Reprinted with permission from *Science*, New Series, Vol. 162, No. 3859 (December 13, 1968), 1243–1248. Copyright © 1968, American Association for the Advancement of Science (AAAS).

implicit and almost universal assumption of discussions published in professional and semipopular scientific journals is that the problem under discussion has a technical solution. A technical solution may be defined as one that requires a change only in the techniques of the natural sciences, demanding little or nothing in the way of change in human values or ideas of morality.

In our day (though not in earlier times) technical solutions are always welcome. Because of previous failures in prophecy, it takes courage to assert that a desired technical solution is not possible. Wiesner and York exhibited this courage; publishing in a science journal, they insisted that the solution to the problem was not to be found in the natural sciences. They cautiously qualified their statement with the phrase, "It is our considered professional judgment . . ." Whether they were right or not is not the concern of the present article. Rather, the concern here is with the important concept of a class of human problems which can be called "no technical solution problems," and, more specifically, with the identification and discussion of one of these.

It is easy to show that the class is not a null class. Recall the game of tick-tack-toe. Consider the problem, "How can I win the game of tick-tack-toe?" It is well known that I cannot, if I assume (in keeping with the conventions of game theory) that my opponent understands the game perfectly. Put another way, there is no "technical solution" to the problem. I can win only by giving a radical meaning to the word "win." I can hit my opponent over the head; or I can drug him; or I can falsify the records. Every way in which I "win" involves, in some sense, an abandonment of the game, as we intuitively understand it. (I can also, of course, openly abandon the game—refuse to play it. This is what most adults do.)

The class of "No technical solution problems" has members. My thesis is that the "population problem," as conventionally conceived, is a member of this class. How it is conventionally conceived needs some comment. It is fair to say that most people who anguish over the population problem are trying to find a way to avoid the evils of overpopulation without relinquishing any of the privileges they now enjoy. They think that farming the seas or developing new strains of wheat will solve the problem—technologically. I try to show here that the solution they seek cannot be found. The population problem cannot be solved in a technical way, any more than can the problem of winning the game of tick-tack-toe.

What Shall We Maximize?

Population, as Malthus said, naturally tends to grow "geometrically," or, as we would now say, exponentially. In a finite world this means that the per capita share of the world's goods must steadily decrease. Is ours a finite world?

A fair defense can be put forward for the view that the world is infinite; or that we do not know that it is not. But, in terms of the practical problems that we must face in the next few generations with the foreseeable technology, it is clear that we will greatly increase human misery if we do not, during the immediate future, assume that the world available to the terrestrial human population is finite. "Space" is no escape (2).

A finite world can support only a finite population; therefore, population growth must eventually equal zero. (The case of perpetual wide fluctuations above and below zero is a trivial variant that need not be discussed.) When this condition is met, what will be the situation of mankind? Specifically, can Bentham's goal of "the greatest good for the greatest number" be realized?

No—for two reasons, each sufficient by itself. The first is a theoretical one. It is not mathematically possible to maximize for two (or more) variables at the same time. This was clearly stated by von Neumann and Morgenstern (3), but the principle is implicit in the theory of partial differential equations, dating back at least to D'Alembert (1717–1783).

The second reason springs directly from biological facts. To live, any organism must have a source of energy (for example, food). This energy is utilized for two purposes: mere maintenance and work. For man, maintenance of life requires about 1600 kilo-calories a day ("maintenance calories"). Anything that he does over and above merely staying alive will be defined as work, and is supported by "work calories" which he takes in. Work calories are used not only for what we call work in common speech; they are also required for all forms of enjoyment, from swimming and automobile racing to playing music and writing poetry. If our goal is to maximize population it is obvious what we must do: We must make the work calories per person approach as close to zero as possible. No gourmet meals, no vacations, no sports, no music, no literature, no art. . . . I think that everyone will grant, without argument or proof, that maximizing population does not maximize goods. Bentham's goal is impossible.

In reaching this conclusion I have made the usual assumption that it is the acquisition of energy that is the problem. The appearance of atomic energy has led some to question this assumption. However, given an infinite source of energy, population growth still produces an inescapable problem. The problem of the acquisition of energy is replaced by the problem of its dissipation, as J. H. Fremlin has so wittily shown (4). The arithmetic signs in the analysis are, as it were, reversed; but Bentham's goat is still unobtainable.

The optimum population is, then, less than the maximum. The difficulty of defining the optimum is enormous; so far as I know, no one has seriously tackled this problem. Reaching an acceptable and stable solution will surely require more than one generation of hard analytical work—and much persuasion.

We want the maximum good per person; but what is good? To one person it is wilderness, to another it is ski lodges for thousands. To one it is estuaries to nourish ducks for hunters to shoot; to another it is factory land. Comparing one good with another is, we usually say, impossible because goods are incommensurable. Incommensurables cannot be compared.

Theoretically this may be true; but in real life incommensurables *are* commensurable. Only a criterion of judgment and a system of weighting are needed. In nature the criterion is survival. Is it better for a species to be small and hide-able, or large and powerful? Natural selection commensurates the incommensurables. The compromise achieved depends on a natural weighting of the values of the variables.

Man must imitate this process. There is no doubt that in fact he already does, but unconsciously. It is when the hidden decisions are made explicit that the arguments begin. The problem for the years ahead is to work out an acceptable theory of weighting. Synergistic effects, nonlinear variation, and difficulties in discounting the future make the intellectual problem difficult, but not (in principle) insoluble.

Has any cultural group solved this practical problem at the present time, even on an intuitive level? One simple fact proves that none has: there is no prosperous population in the world today that has, and has had for some time, a growth rate of zero. Any people that has intuitively identified its optimum point will soon reach it, after which its growth rate becomes and remains zero.

Of course, a positive growth rate might be taken as evidence that a population is below its optimum. However, by any reasonable standards, the most rapidly growing populations on earth today are (in general) the most miserable. This association (which need not be invariable) casts doubt on the optimistic assumption that the positive growth rate of a population is evidence that it has yet to reach its optimum.

We can make little progress in working toward optimum population size until we explicitly exorcize the spirit of Adam Smith in the field of practical demography. In economic affairs, *The Wealth of Nations* (1776) popularized the "invisible hand," the idea that an individual who "intends only his own gain." is, as it were, "led by an invisible hand to promote . . . the public interest" (5). Adam Smith did not assert that this was invariably true, and perhaps neither did any of his followers. But he contributed to a dominant tendency of thought that has ever since interfered with positive action based on rational analysis, namely, the tendency to assume that decisions reached individually will, in fact, be the best decisions for an entire society. If this assumption is correct it justifies the continuance of our present policy of laissez-faire in reproduction. If it is correct we can assume that men will control their individual fecundity so as to produce the optimum population. If the assumption is not correct, we need to reexamine our individual freedoms to see which ones are defensible.

Tragedy of Freedom in a Commons

The rebuttal to the invisible hand in population control is to be found in a scenario first sketched in a little-known pamphlet (6) in 1833 by a mathematical amateur named William Forster Lloyd (1794–1852). We may well call it "the tragedy of the commons," using the word "tragedy" as the philosopher Whitehead used it (7): "The essence of dramatic tragedy is not unhappiness. It resides in the solemnity of the remorseless working of things." He then goes on to say, "This inevitableness of destiny can only be illustrated in terms of human life by incidents which in fact involve unhappiness. For it is only by them that the futility of escape can be made evident in the drama."

The tragedy of the commons develops in this way. Picture a pasture open to all. It is to be expected that each herdsman will try to keep as many cattle as possible on the commons. Such an arrangement may work reasonably satisfactorily for centuries because tribal wars, poaching, and disease keep the numbers of both man and beast well below the carrying capacity of the land. Finally, however, comes the day of reckoning, that is, the day when the long-desired goal of social stability becomes a reality. At this point, the inherent logic of the commons remorselessly generates tragedy.

As a rational being, each herdsman seeks to maximize his gain. Explicitly or implicitly, more or less consciously, he asks, "What is the utility *to me* of adding one more animal to my herd?" This utility has one negative and one positive component.

(1) The positive component is a function of the increment of one animal. Since the herdsman receives all the proceeds from the sale of the additional animal, the positive utility is nearly +1.
(2) The negative component is a function of the additional overgrazing created by one more animal. Since, however, the effects of overgrazing are shared by all the herdsmen, the negative utility for any particular decision-making herdsman is only a fraction of −1.

Adding together the component partial utilities, the rational herdsman concludes that the only sensible course for him to pursue is to add another animal to his herd. And another; and another. . . . But this is the conclusion reached by each and every rational herdsman sharing a commons. Therein is the tragedy. Each man is locked into a system that compels him to increase his herd without limit—in a world that is limited. Ruin is the destination toward which all men rush, each pursuing his own best interest in a society that believes in the freedom of the commons. Freedom in a commons brings ruin to all.

Some would say that this is a platitude. Would that it were! In a sense, it was learned thousands of years ago, but natural selection favors the forces of psychological denial (8). The individual benefits as an individual from his ability to deny the truth even though society as a whole, of which he is a part, suffers. Education can counteract the natural tendency to do the wrong thing, but the inexorable succession of generations requires that the basis for this knowledge be constantly refreshed.

A simple incident that occurred a few years ago in Leominster, Massachusetts, shows how perishable the knowledge is. During the Christmas shopping season the parking meters downtown were covered with plastic bags that bore tags reading: "Do not open until after Christmas. Free parking courtesy of the mayor and city council." In other words, facing the prospect of an increased demand for already scarce space, the city fathers reinstituted the system of the commons. (Cynically, we suspect that they gained more votes than they lost by this retrogressive act.)

In an approximate way, the logic of the commons has been understood for a long time, perhaps since the discovery of agriculture or the invention of private property in real estate. But it is understood mostly only in special cases which are not sufficiently generalized. Even at this late date, cattlemen leasing national land on the western ranges demonstrate no more than an ambivalent understanding, in constantly pressuring federal authorities to increase the head count to the point where overgrazing produces erosion and weed-dominance. Likewise, the oceans of the world continue to suffer from the survival of the philosophy of the commons. Maritime nations still respond automatically to the shibboleth of the "freedom of the seas." Professing to believe in the "inexhaustible resources of the oceans," they bring species after species of fish and whales closer to extinction (9).

The National Parks present another instance of the working out of the tragedy of the commons. At present, they are open to all, without limit. The parks themselves are limited in extent—there is only one Yosemite Valley—whereas population seems to grow without limit. The values that visitors seek in the parks are steadily eroded. Plainly, we must soon cease to treat the parks as commons or they will be of no value to anyone.

What shall we do? We have several options. We might sell them off as private property. We might keep them as public property, but allocate the right to enter them. The allocation might be on the basis of wealth, by the use of an auction system. It might be on the basis of merit, as defined by some agreed-upon standards, it might be by lottery. Or it might be on a first-come, first-served basis, administered to long queues. These, I think, are all the reasonable possibilities. They are all objectionable. But we must choose—or acquiesce in the destruction of the commons that we call our National Parks.

POLLUTION

In a reverse way, the tragedy of the commons reappears in problems of pollution. Here it is not a question of taking something out of the commons, but of putting something in—sewage, or chemical, radioactive, and heat wastes into water; noxious and dangerous fumes into the air; and distracting and unpleasant advertising signs into the line of sight. The calculations of utility are much the same as before. The rational man finds that his share of the cost of the wastes he discharges into the commons is less than the cost of purifying his wastes before releasing them. Since this is true for everyone, we are locked into a system of "fouling our own nest," so long as we behave only as independent, rational, free-enterprisers.

The tragedy of the commons as a food basket is averted by private property, or something formally like it. But the air and waters surrounding us cannot readily be fenced, and so the tragedy of the commons as a cesspool must be prevented by different means, by coercive laws or taxing devices that make it cheaper for the polluter to treat his pollutants than to discharge them untreated. We have not progressed as far with the solution of this problem as we have with the first. Indeed, our particular concept of private property, which deters us from exhausting the positive resources of the earth, favors pollution. The owner of a factory on the bank of a stream—whose property extends to the middle of the stream—often has difficulty seeing why it is not his natural right to muddy the waters flowing past his door. The law, always behind the times, requires elaborate stitching and fitting to adapt it to this newly perceived aspect of the commons.

The pollution problem is a consequence of population. It did not much matter how a lonely American frontiersman disposed of his waste. "Flowing water purifies itself every 10 miles," my grandfather used to say, and the myth was near enough to the truth when he was a boy, for there were not too many people. But as population became denser, the natural chemical and biological recycling processes became overloaded, calling for a redefinition of property rights.

HOW TO LEGISLATE TEMPERANCE?

Analysis of the pollution problem as a function of population density uncovers a not generally recognized principle of morality, namely: *the morality of an act is a junction of the state of the system at the time it is performed* (10). Using the commons as a cesspool does not harm the general public under frontier conditions, because there is no public; the same behavior in a metropolis is unbearable. A hundred and fifty years ago a plainsman could kill an American bison, cut out only the tongue for his dinner, and discard the rest of the animal. He was not in any important sense being wasteful. Today, with only a few thousand bison left, we would be appalled at such behavior.

In passing, it is worth noting that the morality of an act cannot be determined from a photograph. One does not know whether a man killing an elephant or setting fire to the grassland is harming others until one knows the total system in which his act appears. "One picture is worth a thousand words," said an ancient Chinese; but it may take 10,000 words to validate it. It is as tempting to ecologists as it is to reformers in general to try to persuade others by way of the photographic shortcut But the essense of an argument cannot be photographed: it must be presented rationally—in words.

That morality is system-sensitive escaped the attention of most codifiers of ethics in the past. "Thou shalt not . . ." is the form of traditional ethical directives which make no allowance for particular circumstances. The laws of our society follow the pattern of ancient ethics, and therefore are poorly suited to governing a complex, crowded, changeable world. Our epicyclic solution is to augment statutory law with administrative law. Since it is practically impossible to spell out all the conditions under which it is safe to burn trash in the back yard or to run an automobile without smog-control, by law we delegate the details to bureaus. The result is administrative law, which is rightly feared for an ancient reason—*Quis custodiet ipsos custodes?*— "Who shall watch the watchers themselves?" John Adams said that we must have "a government of laws and not men." Bureau administrators, trying to evaluate the morality of acts in the total system, are singularly liable to corruption, producing a government by men, not laws.

Prohibition is easy to legislate (though not necessarily to enforce); but how do we legislate temperance? Experience indicates that it can be accomplished best through the mediation of administrative law. We limit possibilities unnecessarily if we suppose that the sentiment of *Quis custodiet* denies us the use of administrative law. We should rather retain the phrase as a perpetual reminder of fearful dangers we cannot avoid. The great challenge facing us now is to invent the corrective feedbacks that are needed to keep custodians honest. We must find ways to legitimate the needed authority of both the custodians and the corrective feedbacks.

Freedom to Breed is Intolerable

The tragedy of the commons is involved in population problems in another way. In a world governed solely by the principle of "dog eat dog"—if indeed there ever was such a world—how many children a family had would not be a matter of public concern. Parents who bred too exuberantly would leave fewer descendants, not more, because they would be unable to care adequately for their children. David Lack and others have found that such a negative feedback demonstrably controls the fecundity of birds (11). But men are not birds, and have not acted like them for millenniums, at least.

If each human family were dependent only on its own resources; *if* the children of improvident parents starved to death; *if*, thus, overbreeding brought its own "punishment" to the germ line—*then* there would be no public interest in controlling the breeding of families. But our society is deeply committed to the welfare state (12), and hence is confronted with another aspect of the tragedy of the commons.

In a welfare state, how shall we deal with the family, the religion, the race, or the class (or indeed any distinguishable and cohesive group) that adopts overbreeding as a policy to secure its own aggrandizement (13)? To couple the concept of freedom to breed with the belief that everyone born has an equal right to the commons is to lock the world into a tragic course of action.

Unfortunately this is just the course of action that is being pursued by the United Nations. In late 1967, some 30 nations agreed to the following (14):

> The Universal Declaration of Human Rights describes the family as the natural and fundamental unit of society. It follows that any choice and decision with regard to the size of the family must irrevocably rest with the family itself, and cannot be made by anyone else.

It is painful to have to deny categorically the validity of this right; denying it, one feels as uncomfortable as a resident of Salem, Massachusetts, who denied the reality of witches in the 17th century. At the present time, in liberal quarters, something like a taboo acts to inhibit criticism of the United

Nations. There is a feeling that the United Nations is "our last and best hope," that we shouldn't find fault with it; we shouldn't play into the hands of the archconservatives. However, let us not forget what Robert Louis Stevenson said: "The truth that is suppressed by friends is the readiest weapon of the enemy." If we love the truth we must openly deny the validity of the Universal Declaration of Human Rights, even though it is promoted by the United Nations. We should also join with Kingsley Davis (15) in attempting to get Planned Parenthood-World Population to see the error of its ways in embracing the same tragic ideal.

CONSCIENCE IS SELF-ELIMINATING

It is a mistake to think that we can control the breeding of mankind in the long run by an appeal to conscience. Charles Galton Darwin made this point when he spoke on the centennial of the publication of his grandfather's great book. The argument is straightforward and Darwinian.

People vary. Confronted with appeals to limit breeding, some people will undoubtedly respond to the plea more than others. Those who have more children will produce a larger fraction of the next generation than those with more susceptible consciences. The difference will be accentuated, generation by generation.

In C. G. Darwin's words: "It may well be that it would take hundreds of generations for the progenitive instinct to develop in this way, but if it should do so, nature would have taken her revenge, and the variety *Homo contracipiens* would become extinct and would be replaced by the variety *Homo progenitivus*" (16).

The argument assumes that conscience or the desire for children (no matter which) is hereditary—but hereditary only in the most general formal sense. The result will be the same whether the attitude is transmitted through germ cells, or exosomatically, to use A. J. Lotka's term. (If one denies the latter possibility as well as the former, then what's the point of education?) The argument has here been stated in the context of the population problem, but it applies equally well to any instance in which society appeals to an individual exploiting a commons to restrain himself for the general good—by means of his conscience. To make such an appeal is to set up a selective system that works toward the elimination of conscience from the race.

PATHOGENIC EFFECTS OF CONSCIENCE

The long-term disadvantage of an appeal to conscience should be enough to condemn it; but has serious short-term disadvantages as well. If we ask a man who is exploiting a commons to desist "in the name of conscience," what are we saying to him? What does he hear?—not only at the moment but also in the wee small hours of the night when, half asleep, he remembers not merely the words we used but also the nonverbal communication cues we gave him unawares? Sooner or later, consciously or subconsciously, he senses that he has received two communications, and that they are contradictory: (i) (intended communication) "If you don't do as we ask, we will openly condemn you for not acting like a responsible citizen"; (ii) (the unintended communication) "If you *do* behave as we ask, we will secretly condemn you for a simpleton who can be shamed into standing aside while the rest of us exploit the commons."

Everyman then is caught in what Bateson has called a "double bind." Bateson and his co-workers have made a plausible case for viewing the double bind as an important causative factor in the genesis of schizophrenia (17). The double bind may not always be so damaging, but it always endangers the mental health of anyone to whom it is applied. "A bad conscience," said Nietzsche, "is a kind of illness."

To conjure up a conscience in others is tempting to anyone who wishes to extend his control beyond the legal limits. Leaders at the highest level succumb to this temptation. Has any President during the past generation failed to call on labor unions to moderate voluntarily their demands for higher wages, or to steel companies to honor voluntary guidelines on prices? I can recall none. The rhetoric used on such occasions is designed to produce feelings of guilt in noncooperators.

For centuries it was assumed without proof that guilt was a valuable, perhaps even an indispensable, ingredient of the civilized life. Now, in this post-Freudian world, we doubt it.

Paul Goodman speaks from the modern point of view when he says: "No good has ever come from feeling guilty, neither intelligence, policy, nor compassion. The guilty do not pay attention to the object but only to themselves, and not even to their own interests, which might make sense, but to their anxieties" (18).

One does not have to be a professional psychiatrist to see the consequences of anxiety. We in the Western world are just emerging from a dreadful two-centuries-long Dark Ages of Eros that was sustained partly by prohibition laws, but perhaps more effectively by the anxiety-generating mechanisms of education. Alex Comfort has told the story well in *The Anxiety Makers* (19); it is not a pretty one.

Since proof is difficult, we may even concede that the results of anxiety may sometimes, from certain points of view, be desirable. The larger question we should ask is whether, as a matter of policy, we should ever encourage the use of a technique the tendency (if not the intention) of which is psychologically pathogenic. We hear much talk these days of responsible parenthood; the coupled words are incorporated into the titles of some organizations devoted to birth control. Some people have proposed massive propaganda campaigns to instill responsibility into the nation's (or the world's) breeders. But what is the meaning of the word responsibility in this context? Is it not merely a synonym for the word conscience? When we use the word responsibility in the absence of substantial sanctions are we not trying to browbeat a free man in a commons into acting against his own interest? Responsibility is a verbal counterfeit for a substantial *quid pro quo*. It is an attempt to get something for nothing.

If the word responsibility is to be used at all, I suggest that it be in the sense Charles Frankel uses it (20). "Responsibility," says this philosopher, "is the product of definite social arrangements." Notice that Frankel calls for social arrangements—not propaganda.

Mutual Coercion Mutually Agreed upon

The social arrangements that produce responsibility are arrangements that create coercion, of some sort. Consider bank-robbing. The man who takes money from a bank acts as if the bank were a commons. How do we prevent such action? Certainly not by trying to control his behavior solely by a verbal appeal to his sense of responsibility. Rather than rely on propaganda we follow Frankel's lead and insist that a bank is not a commons; we seek the definite social arrangements that will keep it from becoming a commons. That we thereby infringe on the freedom of would-be robbers we neither deny nor regret.

The morality of bank-robbing is particularly easy to understand because we accept complete prohibition of this activity. We are willing to say "Thou shalt not rob banks," without providing for exceptions. But temperance also can be created by coercion. Taxing is a good coercive device. To keep downtown shoppers temperate in their use of parking space we introduce parking meters for short periods, and traffic fines for longer ones. We need not actually forbid a citizen to park as long as he wants to; we need merely make it increasingly expensive for him to do so. Not prohibition, but carefully biased options are what we offer him. A Madison Avenue man might call this persuasion; I prefer the greater candor of the word coercion.

Coercion is a dirty word to most liberals now, but it need not forever be so. As with the four-letter words, its dirtiness can be cleansed away by exposure to the light, by saying it over and over without apology or embarrassment. To many, the word coercion implies arbitrary decisions of distant and irresponsible bureaucrats; but this is not a necessary part of its meaning. The only kind of coercion I recommend is mutual coercion, mutually agreed upon by the majority of the people affected.

To say that we mutually agree to coercion is not to say that we are required to enjoy it, or even to pretend we enjoy it. Who enjoys taxes? We all grumble about them. But we accept compulsory taxes because we recognize that voluntary taxes would favor the conscienceless. We institute and (grumblingly) support taxes and other coercive devices to escape the horror of the commons.

An alternative to the commons need not be perfectly just to be preferable. With real estate and other material goods, the alternative we have chosen is the institution of private property coupled with legal inheritance. Is this system perfectly just? As a genetically trained biologist I deny that it is. It seems to me that, if there are to be differences in individual inheritance, legal possession should be perfectly correlated with biological inheritance—that those who are biologically more fit to be the custodians of property and power should legally inherit more. But genetic recombination continually makes a mockery of the doctrine of "like father, like son" implicit in our laws of legal inheritance. An idiot can inherit millions, and a trust fund can keep his estate intact. We must admit that our legal system of private property plus inheritance is unjust—but we put up with it because we are not convinced, at the moment, that anyone has invented a better system. The alternative of the commons is too horrifying to contemplate. Injustice is preferable to total ruin.

It is one of the peculiarities of the warfare between reform and the status quo that it is thoughtlessly governed by a double standard. Whenever a reform measure is proposed it is often defeated when its opponents triumphantly discover a flaw in it. As Kingsley Davis has pointed out (21), worshippers of the status quo sometimes imply that no reform is possible without unanimous agreement, an implication contrary to historical fact. As nearly as I can make out, automatic rejection of proposed reforms is based on one of two unconscious assumptions: (i) that the status quo is perfect; or (ii) that the choice we face is between reform and no action; if the proposed reform is imperfect, we presumably should take no action at all, while we wait for a perfect proposal.

But we can never do nothing. That which we have done for thousands of years is also action. It also produces evils. Once we are aware that the status quo is action, we can then compare its discoverable advantages and disadvantages with the predicted advantages and disadvantages of the proposed reform, discounting as best we can for our lack of experience. On the basis of such a comparison, we can make a rational decision which will not involve the unworkable assumption that only perfect systems are tolerable.

Recognition of Necessity

Perhaps the simplest summary of this analysis of man's population problems is this: the commons, if justifiable at all, is justifiable only under conditions of low-population density. As the human population has increased, the commons has had to be abandoned in one aspect after another.

First we abandoned the commons in food gathering, enclosing farm land and restricting pastures and hunting and fishing areas. These restrictions are still not complete throughout the world.

Somewhat later we saw that the commons as a place for waste disposal would also have to be abandoned. Restrictions on the disposal of domestic sewage are widely accepted in the Western world; we are still struggling to close the commons to pollution by automobiles, factories, insecticide sprayers, fertilizing operations, and atomic energy installations.

In a still more embryonic state is our recognition of the evils of the commons in matters of pleasure. There is almost no restriction on the propagation of sound waves in the public medium. The shopping public is assaulted with mindless music, without its consent. Our government is paying out billions of dollars to create supersonic transport which will disturb 50,000 people for every one person who is whisked from coast to coast 3 hours faster. Advertisers muddy the airwaves of radio and television and pollute the view of travelers. We are a long way from outlawing the commons in matters of pleasure. Is this because our Puritan inheritance makes us view pleasure as something of a sin, and pain (that is, the pollution of advertising) as the sign of virtue?

Every new enclosure of the commons involves the infringement of somebody's personal liberty. Infringements made in the distant past are accepted because no contemporary complains of a loss, it is the newly proposed infringements that we vigorously oppose: cries of "rights" and "freedom" fill the air. But what does "freedom" mean? When men mutually agreed to pass laws against robbing, mankind became more free, not less so. Individuals locked into the logic of the commons are free only to bring on universal ruin; once they see the necessity of mutual coercion, they become free to pursue other goals. I believe it was Hegel who said, "Freedom is the recognition of necessity."

The most important aspect of necessity that we must now recognize, is the necessity of abandoning the commons in breeding. No technical solution can rescue us from the misery of overpopulation. Freedom to breed will bring ruin to all. At the moment, to avoid hard decisions many of us are tempted to propagandize for conscience and responsible parenthood. The temptation must be resisted because an appeal to independently acting consciences selects for the disappearance of all conscience in the long run, and an increase in anxiety in the short.

The only way we can preserve and nurture other and more precious freedoms is by relinquishing the freedom to breed, and that very soon. "Freedom is the recognition of necessity"—and it is the role of education to reveal to all the necessity of abandoning the freedom to breed. Only so, can we put an end to this aspect of the tragedy of the commons.

REFERENCES

1. J. B. Wiesner and H. F. York, *Sci. Amer.* 211 (No. 4), 27 (1964).
2. G. Hardin, *J. Hered.* 50, 68 (1959); S. von Hoernor, *Science* 137, 18 (1962).
3. J. von Neumann and O. Morgenstern, *Theory of Games and Economic Behavior* (Princeton Univ. Press, Princeton, NJ, 1947), p. 11.
4. J. H. Fremlin, *New Sci.* No. 415 (1968), p. 285.
5. A. Smith, *The Wealth of Nations* (Modern Library, New York, 1937), p. 423.
6. W. F. Lloyd, *Two Lectures on the Checks to Population* (Oxford Univ. Press, Oxford, England, 1833), reprinted (in part) in *Population, Evolution, and Birth Control*, G. Hardin, Ed. (Freeman, San Francisco, 1964), p. 37.
7. A. N. Whitehead, *Science and the Modern World* (Mentor, New York, 1948), p. 17.
8. G. Hardin, Ed., *Population, Evolution, and Birth Control* (Freeman, San Francisco, 1964), p. 56.
9. S. McVay, *Sci. Amer.* 216 (No. 8), 13 (1966).
10. J. Fletcher, *Situation Ethics* (Westminster, Philadelphia, 1966).
11. B. Lack, *The Natural Regulation of Animal Numbers* (Clarendon Press, Oxford, 1954).

12. H. Girvetz, *From Wealth to Welfare* (Stanford Univ. Press, Stanford, CA, 1950).
13. G. Hardin, *Perspec. Biol. Med.* 6, 366 (1963).
14. U. Thant, *Int. Planned Parenthood News* No. 168 (February 1968), p. 3.
15. K. Davis, *Science* 158, 730 (1967).
16. S. Tax, Ed., *Evolution after Darwin* (Univ. of Chicago Press, Chicago, 1960), vol. 2, p. 469.
17. G. Bateson, D. D. Jackson, J. Haley, J. Weakland, *Behav. Sci.* 1, 251 (1956).
18. P. Goodman, *New York Rev. Books* 10 (8), 22 (May 23, 1968).
19. A. Comfort, *The Anxiety Makers* (Nelson, London, 1967).
20. C. Frankel, *The Case for Modern Man* (Harper, New York, 1955), p. 203.
21. J. D. Roslassky, *Genetics and the Future of Man* (Appleton-Century Crofts, New York, 1966), p. 177.

24

Sierra Club v. Morton (1972)

In 1965 the Sierra Club brought suit against the United State Department of the Interior to protect Mineral King valley in the Sierra Nevada Mountains of California from the development of a planned ski resort by the Walt Disney Corporation. As the case progressed, the chief issue in controversy proved to be the preliminary question of whether the Sierra Club had standing to sue. Rather than alleging that the club or its individual members had a specific interest in Mineral King arising from their use of the valley, the petitioners sued on behalf of the valley, claiming that the Sierra Club had "a special interest in the conservation and sound maintenance of the national parks, game refuges, and forests of the country," and therefore should be given standing to intervene on behalf of the true party in interest—the Mineral King valley itself. The District Court issued a preliminary injunction against the development, but this decision was reversed by the Court of Appeals, which held that the Sierra Club lacked legal standing to maintain the action because it had not alleged individualized harm to itself or its members.

The case was appealed to the U.S. Supreme Court and argued before the full court on November 17, 1971. In the period between the hearing of the case and the court handing down its decision on April 19, 1972, this issue of whether a natural object should itself have legal standing in the courts became a hot issue when a young legal scholar, Christopher D. Stone, published an article in the Southern California Law Review *entitled "Should Trees Have Standing? Toward Legal Rights for Natural Objects" (1972). Justice William O. Douglas cited Stone's article in his dissent, referring to other instances in which inanimate objects, such as ships or corporations, had been granted legal standing. The suit, wrote Douglas, "should therefore be more properly labeled as Mineral King v. Morton." A narrowly split court rejected this line of reasoning, but broadly hinted to the Sierra Club that if the suit was redrafted to indicate a personal interest in the valley, the case would be considered on the merits—a tactic that was indeed pursued on appeal. While a precedent for granting legal standing to natural objects was not established by the court in Sierra Club v. Morton, the case inspired numerous legal scholars and environmentalists to revisit the issue of whether legal rights and protections should (or could) be extended to cover nonhuman plaintiffs, such as animals or ecosystems.*

U.S. Supreme Court
Sierra Club v. Morton, 405 U.S. 727 (1972)
Sierra Club v. Morton, Secretary of the Interior, et al.
Certiorari to the United States Court of Appeals for the Ninth Circuit
No. 70-34.
Argued November 17, 1971
Decided April 19, 1972

Petitioner, a membership corporation with "a special interest in the conservation and sound maintenance of the national parks, game refuges, and forests of the country," brought this suit for a declaratory judgment and an injunction to restrain federal officials from approving an extensive skiing development in the Mineral King Valley in the Sequoia National Forest. Petitioner relies on [§] 10 of the Administrative Procedure Act, which accords judicial review to a "person suffering legal wrong because of agency action, or [who is] adversely affected or aggrieved by agency action within the meaning of a relevant statute." On the theory that this was a "public" action involving questions as to the use of natural resources, petitioner did not allege that the challenged development would affect the club or its members in their activities or that they used Mineral King, but maintained that the project would adversely change the area's aesthetics and ecology. The District Court granted a preliminary injunction. The Court of Appeals reversed, holding that the club lacked standing, and had not shown irreparable injury. Held: A person has standing to seek judicial review under the Administrative Procedure Act only if he can show that he himself has suffered or will suffer injury, whether economic or otherwise. In this case, where petitioner asserted no individualized harm to itself or its members, it lacked standing to maintain the action. Pp. 731–741.

433 F.2d 24, affirmed.

STEWART, J., delivered the opinion of the Court, in which BURGER, C. J., and WHITE and MARSHALL, JJ., joined. DOUGLAS, J., post, p. 741, BRENNAN, J., post, p. 755, and BLACKMUN, J., post, p. 755, filed dissenting opinions. POWELL and REHNQUIST, JJ., took no part in the consideration or decision of the case.

Leland R. Selna, Jr., argued the cause for petitioner. With him on the briefs was Matthew P. Mitchell.

Solicitor General Griswold argued the cause for respondents. With him on the brief were Assistant Attorney General Kashiwa, Deputy Assistant Attorney General Kiechel, William Terry Bray, Edmund B. Clark, and Jacques B. Gelin.

Briefs of amici curiae urging reversal were filed by Anthony A. Lapham and Edward Lee Rogers for the Environmental Defense Fund; by George J. Alexander and Marcel B. Poche for the National Environmental Law

Society; and by Bruce J. Terris and James W. Moorman for the Wilderness Society et al.

Briefs of amici curiae urging affirmance were filed by E. Lewis Reid and Calvin E. Baldwin for the County of Tulare; by Robert C. Keck for the American National Cattlemen's Assn. et al.; and by Donald R. Allen for the Far West Ski Assn. et al. MR. JUSTICE STEWART delivered the opinion of the Court.

I

The Mineral King Valley is an area of great natural beauty nestled in the Sierra Nevada Mountains in Tulare County, California, adjacent to Sequoia National Park. It has been part of the Sequoia National Forest since 1926, and is designated as a national game refuge by special Act of Congress.[1] Though once the site of extensive mining activity, Mineral King is now used almost exclusively for recreational purposes. Its relative inaccessibility and lack of development have limited the number of visitors each year, and at the same time have preserved the valley's quality as a quasiwilderness area largely uncluttered by the products of civilization.

The United States Forest Service, which is entrusted with the maintenance and administration of national forests, began in the late 1940's to give consideration to Mineral King as a potential site for recreational development. Prodded by a rapidly increasing demand for skiing facilities, the Forest Service published a prospectus in 1965, inviting bids from private developers for the construction and operation of a ski resort that would also serve as a summer recreation area. The proposal of Walt Disney Enterprises, Inc., was chosen from those of six bidders, and Disney received a three-year permit to conduct surveys and explorations in the valley in connection with its preparation of a complete master plan for the resort.

The final Disney plan, approved by the Forest Service in January 1969, outlines a $35 million complex of motels, restaurants, swimming pools, parking lots, and other structures designed to accommodate 14,000 visitors daily. This complex is to be constructed on 80 acres of the valley floor under a 30-year use permit from the Forest Service. Other facilities, including ski lifts, ski trails, a cog-assisted railway, and utility installations, are to be constructed on the mountain slopes and in other parts of the valley under a revocable special-use permit. To provide access to the resort, the State of California proposes to construct a highway 20 miles in length. A section of this road would traverse Sequoia National Park, as would a proposed high-voltage power line needed to provide electricity for the resort. Both the highway and the power line require the approval of the Department of the Interior, which is entrusted with the preservation and maintenance of the national parks.

Representatives of the Sierra Club, who favor maintaining Mineral King largely in its present state, followed the progress of recreational planning for the valley with close attention and increasing dismay. They unsuccessfully sought a public hearing on the proposed development in 1965, and in subsequent correspondence with officials of the Forest Service and the Department of the Interior, they expressed the Club's objections to Disney's plan as a whole and to particular features included in it. In June 1969 the Club filed the present suit in the United States District Court for the Northern District of California, seeking a declaratory judgment that various aspects of the proposed development contravene federal laws and regulations governing the preservation of national parks, forests, and game refuges,[2] and also seeking preliminary and permanent injunctions restraining the federal officials involved from granting their approval or issuing permits in connection with the Mineral King project. The petitioner Sierra Club sued as a membership corporation with "a special interest in the conservation and the sound maintenance of the national parks, game refuges and forests of the country," and invoked the judicial-review provisions of the Administrative Procedure Act, 5 U.S.C. 701 et seq.

After two days of hearings, the District Court granted the requested preliminary injunction. It rejected the respondents' challenge to the Sierra Club's standing to sue, and determined that the hearing had raised questions "concerning possible excess of statutory authority, sufficiently substantial and serious to justify a preliminary injunction . . ." The respondents appealed, and the Court of Appeals for the Ninth Circuit reversed. 433 F.2d 24. With respect to the petitioner's standing, the court noted that there was "no allegation in the complaint that members of the Sierra Club would be affected by the actions of [the respondents] other than the fact that the actions are personally displeasing or distasteful to them," id., at 33, and concluded:

> "We do not believe such club concern without a showing of more direct interest can constitute standing in the legal sense sufficient to challenge the exercise of responsibilities on behalf of all the citizens by two cabinet level officials of the government acting under Congressional and Constitutional authority." Id., at 30.

Alternatively, the Court of Appeals held that the Sierra Club had not made an adequate showing of irreparable injury and likelihood of success on the merits to justify issuance of a preliminary injunction. The court thus vacated the injunction. The Sierra Club filed a petition for a writ of certiorari which we granted, (401 U.S. 907), to review the questions of federal law presented.

II

The first question presented is whether the Sierra Club has alleged facts that entitle it to obtain judicial review of the challenged action. Whether a party has a sufficient stake in an otherwise justiciable controversy to obtain judicial resolution of that controversy is what has traditionally been referred to as the question of standing to sue. Where the party does not rely on any specific statute authorizing invocation of the judicial process, the question of standing depends upon whether the party has alleged such a "personal stake in the outcome of the controversy," Baker v. Carr, 369 U.S. 186, 204, as to ensure that "the dispute sought to be adjudicated will be presented in an adversary context and in a form historically viewed as capable of judicial resolution." Flast v. Cohen, 392 U.S. 83, 101. Where, however, Congress has authorized public officials to perform certain functions according to law, and has provided by statute for judicial review of those actions under certain circumstances, the inquiry as to standing must begin with a determination of whether the statute in question authorizes review at the behest of the plaintiff.[3]

The Sierra Club relies upon [§] 10 of the Administrative Procedure Act (APA), 5 U.S.C. 702, which provides:

> "A person suffering legal wrong because of agency action, or adversely affected or aggrieved by agency action within the meaning of a relevant statute, is entitled to judicial review thereof."

Early decisions under this statute interpreted the language as adopting the various formulations of "legal interest" and "legal wrong" then prevailing as constitutional requirements of standing.[4] But, in Data Processing Service v. Camp, 397 U.S. 150, and Barlow v. Collins, 397 U.S. 159, decided the same day, we held more broadly that persons had standing to obtain judicial review of federal agency action under 10 of the APA where they had alleged that the challenged action had caused them "injury in fact," and where the alleged injury was to an interest "arguably within the zone of interests to be protected or regulated" by the statutes that the agencies were claimed to have violated.[5]

In Data Processing, the injury claimed by the petitioners consisted of harm to their competitive position in the computer-servicing market through a ruling by the Comptroller of the Currency that national banks might perform data-processing services for their customers. In Barlow, the petitioners were tenant farmers who claimed that certain regulations of the Secretary of Agriculture adversely affected their economic position vis-a-vis their landlords. These palpable economic injuries have long been recognized as sufficient to lay the basis for standing, with or without a specific statutory provision for judicial review.[6] Thus, neither Data Processing nor Barlow addressed itself to the question, which has arisen with increasing frequency in federal courts in recent years, as to what must be alleged by persons who claim injury of a noneconomic nature to interests that are widely shared.[7] That question is presented in this case.

III

The injury alleged by the Sierra Club will be incurred entirely by reason of the change in the uses to which Mineral King will be put, and the attendant change in the aesthetics and ecology of the area. Thus, in referring to the road to be built through Sequoia National Park, the complaint alleged that the development "would destroy or otherwise adversely affect the scenery, natural and historic objects and wildlife of the park and would impair the enjoyment of the park for future generations." We do not question that this type of harm may amount to an "injury in fact" sufficient to lay the basis for standing under [§] 10 of the APA. Aesthetic and environmental well-being, like economic well-being, are important ingredients of the quality of life in our society, and the fact that particular environmental interests are shared by the many rather than the few does not make them less deserving of legal protection through the judicial process. But the "injury in fact" test requires more than an injury to a cognizable interest. It requires that the party seeking review be himself among the injured.

The impact of the proposed changes in the environment of Mineral King will not fall indiscriminately upon every citizen. The alleged injury will be felt directly only by those who use Mineral King and Sequoia National Park, and for whom the aesthetic and recreational values of the area will be lessened by the highway and ski resort. The Sierra Club failed to allege that it or its members would be affected in any of their activities or pastimes by the Disney development. Nowhere in the pleadings or affidavits did the Club state that its members use Mineral King for any purpose, much less that they use it in any way that would be significantly affected by the proposed actions of the respondents.[8]

The Club apparently regarded any allegations of individualized injury as superfluous, on the theory that this was a "public" action involving questions as to the use of natural resources, and that the Club's longstanding concern with and expertise in such matters were sufficient to give it standing as a "representative of the public."[9] This theory reflects a misunderstanding of our cases involving so-called "public actions" in the area of administrative law.

The origin of the theory advanced by the Sierra Club may be traced to a dictum in Scripps-Howard Radio v. FCC, 316 U.S. 4, in which the licensee of a radio station in Cincinnati, Ohio, sought a stay of an order of the FCC allowing another radio station in a nearby city to change its frequency and increase its range. In discussing its power to grant a stay, the Court noted that "these private litigants have standing only as representatives of the public interest." Id., at 14. But that observation did not describe the basis upon which the appellant was allowed to obtain judicial review as a "person aggrieved" within the meaning of the statute involved in that case,[10] since Scripps-Howard was clearly "aggrieved" by reason of the economic injury that it would suffer as a result of the Commission's action.[11] The Court's statement was, rather, directed to the theory upon which Congress had authorized judicial review of the Commission's actions. That theory had been described earlier in FCC v. Sanders Bros. Radio Station, 309 U.S. 470, 477, as follows:

> "Congress had some purpose in enacting 402 (b) (2). It may have been of opinion that one likely to be financially injured by the issue of a license would be the only person having a sufficient interest to bring to the attention of the appellate court errors of law in the action of the Commission in granting the

license. It is within the power of Congress to confer such standing to prosecute an appeal."

Taken together, Sanders and Scripps-Howard thus established a dual proposition: the fact of economic injury is what gives a person standing to seek judicial review under the statute, but once review is properly invoked, that person may argue the public interest in support of his claim that the agency has failed to comply with its statutory mandate.[12] It was in the latter sense that the "standing" of the appellant in Scripps-Howard existed only as a "representative of the public interest." It is in a similar sense that we have used the phrase "private attorney general" to describe the function performed by persons upon whom Congress has conferred the right to seek judicial review of agency action. See Data Processing, supra, at 154.

The trend of cases arising under the APA and other statutes authorizing judicial review of federal agency action has been toward recognizing that injuries other than economic harm are sufficient to bring a person within the meaning of the statutory language, and toward discarding the notion that an injury that is widely shared is ipso facto not an injury sufficient to provide the basis for judicial review.[13] We noted this development with approval in Data Processing, 397 U.S., at 154, in saying that the interest alleged to have been injured "may reflect 'aesthetic, conservational, and recreational' as well as economic values." But broadening the categories of injury that may be alleged in support of standing is a different matter from abandoning the requirement that the party seeking review must himself have suffered an injury.

Some courts have indicated a willingness to take this latter step by conferring standing upon organizations that have demonstrated "an organizational interest in the problem" of environmental or consumer protection. Environmental Defense Fund v. Hardin, 138 U.S. App. D.C. 391, 395, 428 F.2d 1093, 1097.[14] It is clear that an organization whose members are injured may represent those members in a proceeding for judicial review. See, e.g., NAACP v. Button, 371 U.S. 415, 428. But a mere "interest in a problem," no matter how longstanding the interest and no matter how qualified the organization is in evaluating the problem, is not sufficient by itself to render the organization "adversely affected" or "aggrieved" within the meaning of the APA. The Sierra Club is a large and long-established organization, with a historic commitment to the cause of protecting our Nation's natural heritage from man's depredations. But if a "special interest" in this subject were enough to entitle the Sierra Club to commence this litigation, there would appear to be no objective basis upon which to disallow a suit by any other bona fide "special interest" organization, however small or short-lived. And if any group with a bona fide "special interest" could initiate such litigation, it is difficult to perceive why any individual citizen with the same bona fide special interest would not also be entitled to do so.

The requirement that a party seeking review must allege facts showing that he is himself adversely affected does not insulate executive action from judicial review, nor does it prevent any public interests from being protected through the judicial process.[15] It does serve as at least a rough attempt to put the decision as to whether review will be sought in the hands of those who have a direct stake in the outcome. That goal would be undermined were we to construe the APA to authorize judicial review at the behest of organizations or individuals who seek to do no more than vindicate their own value preferences through the judicial process.[16] The principle that the Sierra Club would have us establish in this case would do just that.

As we conclude that the Court of Appeals was correct in its holding that the Sierra Club lacked standing to maintain this action, we do not reach any other questions presented in the petition, and we intimate no view on the merits of the complaint. The judgment is
Affirmed.

MR. JUSTICE POWELL and MR. JUSTICE REHNQUIST took no part in the consideration or decision of this case.

MR. JUSTICE DOUGLAS, dissenting.

I share the views of my Brother BLACKMUN and would reverse the judgment below.

The critical question of "standing"[1] would be simplified and also put neatly in focus if we fashioned a federal rule that allowed environmental issues to be litigated before federal agencies or federal courts in the name of the inanimate object about to be despoiled, defaced, or invaded by roads and bulldozers and where injury is the subject of public outrage. Contemporary public concern for protecting nature's ecological equilibrium should lead to the conferral of standing upon environmental objects to sue for their own preservation. See Stone, Should Trees Have Standing?—Toward Legal Rights for Natural Objects, 45 S. Cal. L. Rev. 450 (1972). This suit would therefore be more properly labeled as Mineral King v. Morton.

Inanimate objects are sometimes parties in litigation. A ship has a legal personality, a fiction found useful for maritime purposes.[2] The corporation sole—a creature of ecclesiastical law—is an acceptable adversary and large fortunes ride on its cases.[3] The ordinary corporation is a "person" for purposes of the adjudicatory processes, whether it represents proprietary, spiritual, aesthetic, or charitable causes.[4]

So it should be as respects valleys, alpine meadows, rivers, lakes, estuaries, beaches, ridges, groves of trees, swampland, or even air that feels the destructive pressures of modern technology and modern life. The river, for example, is the living symbol of all the life it sustains or nourishes—fish, aquatic insects, water ouzels, otter, fisher, deer, elk, bear, and all other animals, including man, who are dependent on it or who enjoy it for its sight, its sound, or its life. The river as plaintiff speaks for the ecological unit of life that is part of it. Those people who have a meaningful relation to that body of water—whether it be a fisherman, a canoeist, a zoologist, or a logger—must be able to speak for the values which the river represents and which are threatened with destruction.

I do not know Mineral King. I have never seen it nor traveled it, though I have seen articles describing its proposed "development"[5] notably Hano, Protectionists vs. recreationists—The Battle of Mineral King, N. Y. Times Mag., Aug. 17, 1969, p. 25; and Browning, Mickey Mouse in the Mountains, Harper's, March 1972, p. 65. The Sierra Club in its complaint alleges that "[o]ne of the principal purposes of the Sierra Club is to protect and conserve the national resources of the Sierra Nevada Mountains." The District Court held that this uncontested allegation made the Sierra Club "sufficiently aggrieved" to have "standing" to sue on behalf of Mineral King.

Mineral King is doubtless like other wonders of the Sierra Nevada such as Tuolumne Meadows and the John Muir Trail. Those who hike it, fish it, hunt it, camp in it, frequent it, or visit it merely to sit in solitude and wonderment are legitimate spokesmen for it, whether they may be few or many. Those who have that intimate relation with the inanimate object about to be injured, polluted, or otherwise despoiled are its legitimate spokesmen.

The Solicitor General, whose views on this subject are in the Appendix to this opinion, takes a wholly different approach. He considers the problem in terms of "government by the Judiciary." With all respect, the problem is to make certain that the inanimate objects, which are the very core of America's beauty, have spokesmen before they are destroyed. It is, of course, true that most of them are under the control of a federal or state agency. The standards given those agencies are usually expressed in terms of the "public interest." Yet "public interest" has so many differing shades of meaning as to be quite meaningless on the environmental front. Congress accordingly has adopted ecological standards in the National Environmental Policy Act of 1969, Pub. L. 91–190, 83 Stat. 852, 42 U.S.C. 4321 et seq., and guidelines for agency action have been provided by the Council on Environmental Quality of which Russell E. Train is Chairman. See 36 Fed. Reg. 7724.

Yet the pressures on agencies for favorable action one way or the other are enormous. The suggestion that Congress can stop action which is undesirable is true in theory; yet even Congress is too remote to give meaningful direction and its machinery is too ponderous to use very often. The federal agencies of which I speak are not venal or corrupt. But they are notoriously under the control of powerful interests who manipulate them through advisory committees, or friendly working relations, or who have that natural affinity with the agency which in time develops between the regulator and the regulated.[6] As early as 1894, Attorney General Olney predicted that regulatory agencies might become "industry-minded," as illustrated by his forecast concerning the Interstate Commerce Commission:

> "The Commission . . . is, or can be made, of great use to the railroads. It satisfies the popular clamor for a government supervision of railroads, at the same time that that supervision is almost entirely nominal. Further, the older such a commission gets to be, the more inclined it will be found to take the business and railroad view of things." M. Josephson, The Politicos 526 (1938).

Years later a court of appeals observed, "the recurring question which has plagued public regulation of industry [is] whether the regulatory agency is unduly oriented toward the interests of the industry it is designed to regulate, rather than the public interest it is designed to protect." Moss v. CAB, 139 U.S. App. D.C. 150, 152, 430 F.2d 891, 893. See also Office of Communication of the United Church of Christ v. FCC, 123 U.S. App. D.C. 328, 337–338, 359 F.2d 994, 1003–1004; Udall v. FPC, 387 U.S. 428; Calvert Cliffs' Coordinating Committee, Inc. v. AEC, 146 U.S. App. D.C. 33, 449 F.2d 1109; Environmental Defense Fund, Inc. v. Ruckelshaus, 142 U.S. App. D.C. 74, 439 F.2d 584; Environmental Defense Fund, Inc. v. HEW, 138 U.S. App. D.C. 381, 428 F.2d 1083; Scenic Hudson Preservation Conf. v. FPC, 354 F.2d 608, 620. But see Jaffe, The Federal Regulatory Agencies In Perspective: Administrative Limitations In A Political Setting, 11 B. C. Ind. & Com. L. Rev. 565 (1970) (labels "industry-mindedness" as "devil" theory).

The Forest Service—one of the federal agencies behind the scheme to despoil Mineral King—has been notorious for its alignment with lumber companies, although its mandate from Congress directs it to consider the various aspects of multiple use in its supervision of the national forests.[7]

The voice of the inanimate object, therefore, should not be stilled. That does not mean that the judiciary takes over the managerial functions from the federal agency. It merely means that before these priceless bits of Americana (such as a valley, an alpine meadow, a river, or a lake) are forever lost or are so transformed as to be reduced to the eventual rubble of our urban environment, the voice of the existing beneficiaries of these environmental wonders should be heard.[8]

Perhaps they will not win. Perhaps the bulldozers of "progress" will plow under all the aesthetic wonders of this beautiful land. That is not the present question. The sole question is, who has standing to be heard?

Those who hike the Appalachian Trail into Sunfish Pond, New Jersey, and camp or sleep there, or run the Allagash in Maine, or climb the Guadalupes in West Texas, or who canoe and portage the Quetico Superior in Minnesota, certainly should have standing to defend those natural wonders before courts or agencies, though they live 3,000 miles away. Those who merely are caught up in environmental news or propaganda and flock to defend these waters or areas may be treated differently. That is why these environmental issues should be tendered by the inanimate object itself. Then there will be assurances that all of the forms of life[9] which it represents will stand before the court—the pileated woodpecker as well as the coyote and bear, the lemmings as well as the trout in the streams. Those inarticulate members of the ecological group cannot speak. But those people who have so frequented the place as to know its values and wonders will be able to speak for the entire ecological community.

Ecology reflects the land ethic; and Aldo Leopold wrote in A Sand Country Almanac 204 (1949), "The land ethic simply enlarges the boundaries of the community to include soils, waters, plants, and animals, or collectively: the land."

That, as I see it, is the issue of "standing" in the present case and controversy.

* * *

MR. JUSTICE BRENNAN, dissenting.

I agree that the Sierra Club has standing for the reasons stated by my Brother BLACKMUN in Alternative No. 2 of his dissent. I therefore would reach the merits. Since the Court does not do so, however, I simply note agreement with my Brother BLACKMUN that the merits are substantial.

MR. JUSTICE BLACKMUN, dissenting.

The Court's opinion is a practical one espousing and adhering to traditional notions of standing as somewhat modernized by Data Processing Service v. Camp, (1970); Barlow v. Collins, 397 U.S. 159 (1970); and Flast v. Cohen, 392 U.S. 83 (1968). If this were an ordinary case, I would join the opinion and the Court's judgment and be quite content.

But this is not ordinary, run-of-the-mill litigation. The case poses—if only we choose to acknowledge and reach them—significant aspects of a wide, growing, and disturbing problem, that is, the Nation's and the world's deteriorating environment with its resulting ecological disturbances. Must our law be so rigid and our procedural concepts so inflexible that we render ourselves helpless when the existing methods and the traditional concepts do not quite fit and do not prove to be entirely adequate for new issues?

The ultimate result of the Court's decision today, I fear, and sadly so, is that the 35.3-million-dollar complex, over 10 times greater than the Forest Service's suggested minimum, will now hastily proceed to completion; that serious opposition to it will recede in discouragement; and that Mineral King, the "area of great natural beauty nestled in the Sierra Nevada Mountains," to use the Court's words, will become defaced, at least in part, and, like so many other areas, will cease to be "uncluttered by the products of civilization."

I believe this will come about because: (1) The District Court, although it accepted standing for the Sierra Club and granted preliminary injunctive relief, was reversed by the Court of Appeals, and this Court now upholds that reversal. (2) With the reversal, interim relief by the District Court is now out of the question and a permanent injunction becomes most unlikely. (3) The Sierra Club may not choose to amend its complaint or, if it does desire to do so, may not, at this late date, be granted permission. (4) The ever-present pressure to get the project under way will mount. (5) Once under way, any prospect of bringing it to a halt will grow dim. Reasons, most of them economic, for not stopping the project will have a tendency to multiply. And the irreparable harm will be largely inflicted in the earlier stages of construction and development.

Rather than pursue the course the Court has chosen to take by its affirmance of the judgment of the Court of Appeals, I would adopt one of two alternatives:

1. I would reverse that judgment and, instead, approve the judgment of the District Court which recognized standing in the Sierra Club and granted preliminary relief. I would be willing to do this on condition that the Sierra Club forthwith amend its complaint to meet the specifications the Court prescribes for standing. If Sierra Club fails or refuses to take that step, so be it; the case will then collapse. But if it does amend, the merits will be before the trial court once again. As the Court, ante, at 730 n. 2, so clearly reveals, the issues on the merits are substantial and deserve resolution. They assay new ground. They are crucial to the future of Mineral King. They raise important ramifications for the quality of the country's public land management. They pose the propriety of the "dual permit" device as a means of avoiding the 80-acre "recreation and resort" limitation imposed by Congress in 16 U.S.C. 497, an issue that apparently has never been litigated, and is clearly substantial in light of the congressional expansion of the limitation in 1956 arguably to put teeth into the old, unrealistic five-acre limitation. In fact, they concern the propriety of the 80-acre permit itself and the consistency of the entire, enormous development with the statutory purposes of the Sequoia Game Refuge, of which the Valley is a part. In the context of this particular development, substantial questions are raised about the use of a national park area for Disney purposes for a new high speed road and a 66,000-volt power line to serve the complex. Lack of compliance with existing administrative regulations is also charged. These issues are not shallow or perfunctory.

2. Alternatively, I would permit an imaginative expansion of our traditional concepts of standing in order to enable an organization such as the Sierra Club, possessed, as it is, of pertinent, bona fide, and well-recognized attributes and purposes in the area of environment, to litigate environmental issues. This incursion upon tradition need not be very extensive. Certainly, it should be no cause for alarm. It is no more progressive than was the decision in Data Processing itself. It need only recognize the interest of one who has a provable, sincere, dedicated, and established status. We need not fear that Pandora's box will be opened or that there will be no limit to the number of those who desire to participate in environmental litigation. The courts will exercise appropriate restraints just as they have exercised them in the past. Who would have suspected 20 years ago that the concepts of standing enunciated in Data Processing and Barlow would be the measure for today? And MR. JUSTICE DOUGLAS, in his eloquent opinion, has imaginatively suggested another means and one, in its own way, with obvious, appropriate, and self-imposed limitations as to standing. As I read what he has written, he makes only one addition to the customary criteria (the existence of a genuine dispute; the assurance of adversariness; and a conviction that the party whose standing is challenged will adequately represent the interests he asserts), that is, that the litigant be one who speaks knowingly for the environmental values he asserts.

I make two passing references:

1. The first relates to the Disney figures presented to us. The complex, the Court notes, will accommodate 14,000 visitors a day (3,100 overnight; some 800 employees; 10 restaurants; 20 ski lifts). The State of California has proposed to build a new road from Hammond to Mineral King. That road, to the extent of 9.2 miles, is to traverse Sequoia National Park. It will have only two lanes, with occasional passing areas, but it will be capable, it is said, of accommodating 700–800 vehicles per hour and a peak of 1,200 per hour. We are told that the State has agreed not to seek any further improvement in road access through the park.

If we assume that the 14,000 daily visitors come by automobile (rather than by helicopter or bus or other known or unknown means) and that each visiting automobile carries four passengers (an assumption, I am sure, that is far too optimistic), those 14,000 visitors will move in 3,500 vehicles. If we confine their movement (as I think we properly may for this mountain area) to 12 hours out of the daily 24, the 3,500 automobiles will pass any given point on the two-lane road at the rate of about 300 per hour. This amounts to five vehicles per minute, or an average of one every 12 seconds. This frequency is further increased to one every six seconds when the necessary return traffic along that same two-lane road is considered. And this does not include service vehicles and employees' cars. Is this the way we perpetuate the wilderness and its beauty, solitude, and quiet?

2. The second relates to the fairly obvious fact that any resident of the Mineral King area—the real "user"—is an unlikely adversary for this Disney-governmental project. He naturally will be inclined to regard the situation as one that should benefit him economically. His fishing or camping or guiding or handyman or general outdoor prowess perhaps will find an early and ready market among the visitors. But that glow of anticipation will be short-lived at best. If he is a true lover of the wilderness—as is likely, or he would not be near Mineral King in the first place—it will not be long before he yearns for the good old days when masses of people—that 14,000 influx per day—and their thus far uncontrollable waste were unknown to Mineral King.

Do we need any further indication and proof that all this means that the area will no longer be one "of great natural beauty" and one "uncluttered by the products of civilization?" Are we to be rendered helpless to consider and evaluate allegations and challenges of this kind because of procedural limitations rooted in traditional concepts of standing? I suspect that this may be the result of today's holding. As the Court points out, ante, at 738–739, other federal tribunals have not felt themselves so confined.[1] I would join those progressive holdings.

The Court chooses to conclude its opinion with a footnote reference to De Tocqueville. In this environmental context I personally prefer the older and particularly pertinent observation and warning of John Donne.[2]

NOTES

Notes to Pages 201–206

1. Act of July 3, 1926, 6, 44 Stat. 821, 16 U.S.C. 688.
2. As analyzed by the District Court, the complaint alleged violations of law falling into four categories. First, it claimed that the special-use permit for construction of the resort exceeded the maximum-acreage limitation placed upon such permits by 16 U.S.C. 497, and that issuance of a "revocable" use permit was beyond the authority of the Forest Service. Second, it challenged the proposed permit for the highway through Sequoia National Park on the grounds that the highway would not serve any of the purposes of the park, in alleged violation of 16 U.S.C. 1, and that it would destroy timber and other natural resources protected by 16 U.S.C. 41 and 43. Third, it claimed that the Forest Service and the Department of the Interior had violated their own regulations by failing to hold adequate public hearings on the proposed project. Finally, the complaint asserted that 16 U.S.C. 45c requires specific congressional authorization of a permit for construction of a power transmission line within the limits of a national park.
3. Congress may not confer jurisdiction on Art. III federal courts to render advisory opinions, Muskrat v. United States, 219 U.S. 346 , or to entertain "friendly" suits, United States v. Johnson, 319 U.S. 302 , or to resolve "political questions," Luther v. Borden, 7 How. 1, because suits of this character are inconsistent with the judicial function under Art. III. But where a dispute is otherwise justiciable, the question whether the litigant is a "proper party to request an adjudication of a particular issue," Flast v. Cohen, 392 U.S. 83, 100 , is one within the power of Congress to determine. Cf. FCC v. Sanders Bros. Radio Station, 309 U.S. 470, 477 ; Flast v. Cohen, supra, at 120 (Harlan, J., dissenting); Associated Industries v. Ickes, 134 F.2d 694, 704. See generally Berger, Standing to Sue in Public Actions: Is it a Constitutional Requirement?, 78 Yale L. J. 816, 837 et seq. (1969); Jaffe, The Citizen as Litigant in Public Actions: The Non-Hohfeldian or Ideological Plaintiff, 116 U. Pa. L. Rev. 1033 (1968).
4. See, e.g., Kansas City Power & Light Co. v. McKay, 96 U.S. App. D.C. 273, 281, 225 F.2d 924, 932; Ove Gustavsson Contracting Co. v. Floete, 278 F.2d 912, 914; Duba v. Schuetzle, 303 F.2d 570, 574. The theory of a "legal interest" is expressed in its extreme form in Alabama Power Co. v. Ickes, 302 U.S. 464, 479–481. See also Tennessee Electric Power Co. v. TVA, 306 U.S. 118, 137–139.
5. In deciding this case we do not reach any questions concerning the meaning of the "zone of interests" test or its possible application to the facts here presented.
6. See, e.g., Hardin v. Kentucky Utilities Co., 390 U.S. 1, 7 ; Chicago v. Atchison, T. & S. F. R. Co., 357 U.S. 77, 83 ; FCC v. Sanders Bros. Radio Station, supra, at 477.
7. No question of standing was raised in Citizens to Preserve Overton Park v. Volpe, 401 U.S. 402. The complaint in that case alleged that the organizational plaintiff represented members who were "residents of Memphis, Tennessee who use Overton Park as a park land and recreation area and who have been active since 1964 in efforts to preserve and protect Overton Park as a park land and recreation area."
8. The only reference in the pleadings to the Sierra Club's interest in the dispute is contained in paragraph 3 of the complaint, which reads in its entirely as follows:

   > "Plaintiff Sierra Club is a non-profit corporation organized and operating under the laws of the State of California, with its principal place of business in San Francisco, California since 1892. Membership of the club

is approximately 78,000 nationally, with approximately 27,000 members residing in the San Francisco Bay Area. For many years the Sierra Club by its activities and conduct has exhibited a special interest in the conservation and the sound maintenance of the national parks, game refuges and forests of the country, regularly serving as a responsible representative of persons similarly interested. One of the principal purposes of the Sierra Club is to protect and conserve the national resources of the Sierra Nevada Mountains. Its interests would be vitally affected by the acts hereinafter described and would be aggrieved by those acts of the defendants as hereinafter more fully appears."

In an amici curiae brief filed in this Court by the Wilderness Society and others, it is asserted that the Sierra Club has conducted regular camping trips into the Mineral King area, and that various members of the Club have used and continue to use the area for recreational purposes. These allegations were not contained in the pleadings, nor were they brought to the attention of the Court of Appeals. Moreover, the Sierra Club in its reply brief specifically declines to rely on its individualized interest, as a basis for standing. See n. 15, infra. Our decision does not, of course, bar the Sierra Club from seeking in the District Court to amend its complaint by a motion under Rule 15, Federal Rules of Civil Procedure.

9. This approach to the question of standing was adopted by the Court of Appeals for the Second Circuit in Citizens Committee for the Hudson Valley v. Volpe, 425 F.2d 97, 105: "We hold, therefore, that the public interest in environmental resources—an interest created by statutes affecting the issuance of this permit—is a legally protected interest affording these plaintiffs, as responsible representatives of the public, standing to obtain judicial review of agency action alleged to be in contravention of that public interest."
10. The statute involved was 402 (b) (2) of the Communications Act of 1934, 48 Stat. 1093.
11. This much is clear from the Scripps-Howard Court's citation of FCC v. Sanders Bros. Radio Station, 309 U.S. 470, in which the basis for standing was the competitive injury that the appellee would have suffered by the licensing of another radio station in its listening area.
12. The distinction between standing to initiate a review proceeding, and standing to assert the rights of the public or of third persons once the proceeding is properly initiated, is discussed in 3 K. Davis, Administrative Law Treatise 22.05–22.07 (1958).
13. See, e.g., Environmental Defense Fund v. Hardin, 138 U.S. App. D.C. 391, 395, 428 F.2d 1093, 1097 (interest in health affected by decision of Secretary of Agriculture refusing to suspend registration of certain pesticides containing DDT); Office of Communication of the United Church of Christ v. FCC, 123 U.S. App. D.C. 328, 339, 359 F.2d 994. 1005 (interest of television viewers in the programing of a local station licensed by the FCC); Scenic Hudson Preservation Conf. v. FPC, 354 F.2d 608, 615–616 (interests in aesthetics, recreation, and orderly community planning affected by FPC licensing of a hydroelectric project); Reade v. Ewing, 205 F.2d 630, 631–632 (interest of consumers of oleomargarine in fair labeling of product regulated by Federal Security Administration); Crowther v. Seaborg, 312 F. Supp. 1205, 1212 (interest in health and safety of persons residing near the site of a proposed atomic blast).
14. See Citizens Committee for the Hudson Valley v. Volpe, n. 9, supra; Environmental Defense Fund, Inc. v. Corps of Engineers, 325 F. Supp. 728,

734–736; Izaak Walton League v. St. Clair, 313 F. Supp. 1312, 1317. See also Scenic Hudson Preservation Conf. v. FPC, supra, at 616: "In order to insure that the Federal Power Commission will adequately protect the public interest in the aesthetic, conservational, and recreational aspects of power development, those who by their activities and conduct have exhibited a special interest in such areas, must be held to be included in the class of 'aggrieved' parties under 313 (b) [of the Federal Power Act]."

In most, if not all, of these cases, at least one party to the proceeding did assert an individualized injury either to himself or, in the case of an organization, to its members.

15. In its reply brief, after noting the fact that it might have chosen to assert individualized injury to itself or to its members as a basis for standing, the Sierra Club states: "The Government seeks to create a 'heads I win, tails you lose' situation in which either the courthouse door is barred for lack of assertion of a private, unique injury or a preliminary injunction is denied on the ground that the litigant has advanced private injury which does not warrant an injunction adverse to a competing public interest. Counsel have shaped their case to avoid this trap."

    The short answer to this contention is that the "trap" does not exist. The test of injury in fact goes only to the question of standing to obtain judicial review. Once this standing is established, the party may assert the interests of the general public in support of his claims for equitable relief. See n. 12 and accompanying text, supra.

16. Every schoolboy may be familiar with Alexis de Tocqueville's famous observation, written in the 1830's, that "[s]carcely any political question arises in the United States that is not resolved, sooner or later, into a judicial question." 1 Democracy in America 280 (1945). Less familiar, however, is de Tocqueville's further observation that judicial review is effective largely because it is not available simply at the behest of a partisan faction, but is exercised only to remedy a particular, concrete injury.

    > "It will be seen, also, that by leaving it to private interest to censure the law, and by intimately uniting the trial of the law with the trial of an individual, legislation is protected from wanton assaults and from the daily aggressions of party spirit. The errors of the legislator are exposed only to meet a real want; and it is always a positive and appreciable fact that must serve as the basis of a prosecution." Id., at 102.

Notes to Pages 206–208

1. See generally Data Processing Service v. Camp, 397 U.S. 150 (1970); Barlow v. Collins, 397 U.S. 159 (1970); Flast v. Cohen, 392 U.S. 83 (1968). See also MR. JUSTICE BRENNAN'S separate opinion in Barlow v. Collins, supra, at 167. The issue of statutory standing aside, no doubt exists that "injury in fact" to "aesthetic" and "conservational" interests is here sufficiently threatened to satisfy the case-or-controversy clause. Data Processing Service v. Camp, supra, at 154.
2. In rem actions brought to adjudicate libelants' interests in vessels are well known in admiralty. G. Gilmore & C. Black, The Law of Admiralty 31 (1957). But admiralty also permits a salvage action to be brought in the name of the rescuing vessel. The Camanche, 8 Wall. 448, 476 (1869). And, in collision litigation, the first-libeled ship may counterclaim in its own name. The Gylfe v. The Trujillo,

209 F.2d 386 (CA2 1954). Our case law has personified vessels:

> "A ship is born when she is launched, and lives so long as her identity is preserved. Prior to her launching she is a mere congeries of wood and iron. . . . In the baptism of launching she receives her name, and from the moment her keel touches the water she is transformed . . . . She acquires a personality of her own." Tucker v. Alexandroff, 183 U.S. 424, 438.

3. At common law, an officeholder, such as a priest or the king, and his successors constituted a corporation sole, a legal entity distinct from the personality which managed it. Rights and duties were deemed to adhere to this device rather than to the office-holder in order to provide continuity after the latter retired. The notion is occasionally revived by American courts. E.g., Reid v. Barry, 93 Fla. 849, 112 So. 846 (1927), discussed in Recent Cases, 12 Minn. L. Rev. 295 (1928), and in Note, 26 Mich. L. Rev. 545 (1928); see generally 1 W. Fletcher, Cyclopedia of the Law of Private Corporations 50–53 (1963); 1 P. Potter, Law of Corporations 27 (1881).
4. Early jurists considered the conventional corporation to be a highly artificial entity. Lord Coke opined that a corporation's creation "rests only in intendment and consideration of the law." Case of Sutton's Hospital. 77 Eng. Rep. 937, 973 (K. B. 1612). Mr. Chief Justice Marshall added that the device is "an artificial being, invisible, intangible, and existing only in contemplation of law." Trustees of Dartmouth College v. Woodward, 4 Wheat, 518, 636 (1819). Today, suits in the names of corporations are taken for granted.
5. Although in the past Mineral King Valley has annually supplied about 70,000 visitor-days of simpler and more rustic forms of recreation—hiking, camping, and skiing (without lifts)—the Forest Service in 1949 and again in 1965 invited developers to submit proposals to "improve" the Valley for resort use. Walt Disney Productions won the competition and transformed the Service's idea into a mammoth project 10 times its originally proposed dimensions. For example, while the Forest Service prospectus called for an investment of at least $3 million and a sleeping capacity of at least 100, Disney will spend $35.3 million and will bed down 3,300 persons by 1978. Disney also plans a nine-level parking structure with two supplemental lots for automobiles, 10 restaurants and 20 ski lifts. The Service's annual license revenue is hitched to Disney's profits. Under Disney's projections, the Valley will be forced to accommodate a tourist population twice as dense as that in Yosemite Valley on a busy day. And, although Disney has bought up much of the private land near the project, another commercial firm plans to transform an adjoining 160-acre parcel into a "piggyback" resort complex, further adding to the volume of human activity the Valley must endure. See generally Note, Mineral King Valley: Who Shall Watch the Watchmen?, 25 Rutgers L. Rev. 103, 107 (1970); Thar's Gold in Those Hills, 206 The Nation 260 (1968). For a general critique of mass recreation enclaves in national forests see Christian Science Monitor, Nov. 22, 1965, p. 5, col. 1 (Western ed.). Michael Frome cautions that the national forests are "fragile" and "deteriorate rapidly with excessive recreation use" because "[t]he trampling effect alone eliminates vegetative growth, creating erosion and water runoff problems. The concentration of people, particularly in horse parties, on excessively steep slopes that follow old Indian or cattle routes, has torn up the landscape of the High Sierras in California and sent tons of wilderness soil washing downstream each year." M. Frome, The Forest Service 69 (1971).

6. The federal budget annually includes about $75 million for underwriting about 1,500 advisory committees attached to various regulatory agencies. These groups are almost exclusively composed of industry representatives appointed by the President or by Cabinet members. Although public members may be on these committees, they are rarely asked to serve. Senator Lee Metcalf warns: "Industry advisory committees exist inside most important federal agencies, and even have offices in some. Legally, their function is purely as kibitzer, but in practice many have become internal lobbies—printing industry handouts in the Government Printing Office with taxpayers' money, and even influencing policies. Industry committees perform the dual function of stopping government from finding out about corporations while at the same time helping corporations get inside information about what government is doing. Sometimes, the same company that sits on an advisory council that obstructs or turns down a government questionnaire is precisely the company which is withholding information the government needs in order to enforce a law." Metcalf, The Vested Oracles: How Industry Regulates Government, 3 The Washington Monthly, July 1971, p. 45. For proceedings conducted by Senator Metcalf exposing these relationships, see Hearings on S. 3067 before the Subcommittee on Intergovernmental Relations of the Senate Committee on Government Operations, 91st Cong., 2d Sess. (1970); Hearings on S. 1637, S. 1964, and S. 2064 before the Subcommittee on Intergovernmental Relations of the Senate Committee on Government Operations, 92d Cong., 1st Sess. (1971).

   The web spun about administrative agencies by industry representatives does not depend, of course, solely upon advisory committees for effectiveness. See Elman, Administrative Reform of the Federal Trade Commission, 59 Geo. L. J. 777, 788 (1971); Johnson, A New Fidelity to the Regulatory Ideal, 59 Geo. L. J. 869, 874, 906 (1971); R. Berkman & K. Viscusi, Damming The West, The Ralph Nader Study Group Report on The Bureau of Reclamation 155 (1971); R. Fellmeth, The Interstate Commerce Omission, The Ralph Nader Study Group Report on the Interstate Commerce Commission and Transportation 15–39 and passim (1970); J. Turner, The Chemical Feast, The Ralph Nader Study Group Report on Food Protection and the Food and Drug Administration passim (1970); Massel, The Regulatory Process, 26 Law & Contemp. Prob. 181, 189 (1961); J. Landis, Report on Regulatory Agencies to the President-Elect 13, 69 (1960).

7. The Forest Reserve Act of 1897, 30 Stat. 35, 16 U.S.C. 551, imposed upon the Secretary of the Interior the duty to "preserve the [national] forests . . . from destruction" by regulating their "occupancy and use." In 1905 these duties and powers were transferred to the Forest Service created within the Department of Agriculture by the Act of Feb. 1, 1905, 33 Stat. 628, 16 U.S.C. 472. The phrase "occupancy and use" has been the cornerstone for the concept of "multiple use" of national forests, that is, the policy that uses other than logging were also to be taken into consideration in managing our 154 national forests. This policy was made more explicit by the Multiple-Use Sustained-Yield Act of 1960, 74 Stat. 215, 16 U.S.C. 528–531, which provides that competing considerations should include outdoor recreation, range, timber, watershed, wildlife, and fish purposes. The Forest Service, influenced by powerful logging interests, has, however, paid only lip service to its multiple-use mandate and has auctioned away millions of timberland acres without considering environmental or conservational interests. The importance of national forests to the construction and logging industries results from the type of lumber grown therein which is well suited to builders' needs. For

example, Western acreage produces Douglas fir (structural support) and ponderosa pine (plywood lamination). In order to preserve the total acreage and so-called "maturity" of timber, the annual size of a Forest Service harvest is supposedly equated with expected yearly reforestation. Nonetheless, yearly cuts have increased from 5.6 billion board feet in 1950 to 13.74 billion in 1971. Forestry professionals challenge the Service's explanation that this harvest increase to 240% is not really overcutting but instead has resulted from its improved management of timberlands. "Improved management," answer the critics, is only a euphemism for exaggerated regrowth forecasts by the Service. N. Y. Times, Nov. 15, 1971, p. 48, col. 1. Recent rises in lumber prices have caused a new round of industry pressure to auction more federally owned timber. See Wagner, Resources Report/Lumbermen, Conservationists Head for New Battle over Government Timber, 3 National J. 657 (1971).

Aside from the issue of how much timber should be cut annually, another crucial question is how lumber should be harvested. Despite much criticism, the Forest Service had adhered to a policy of permitting logging companies to "clearcut" tracts of auctioned acreage. "Clearcutting," somewhat analogous to strip mining, is the indiscriminate and complete shaving from the earth of all trees—regardless of size or age—often across hundreds of contiguous acres.

Of clearcutting, Senator Gale McGee, a leading antagonist of Forest Service policy, complains: "The Forest Service's management policies are wreaking havoc with the environment. Soil is eroding, reforestation is neglected if not ignored, streams are silting, and clearcutting remains a basic practice." N. Y. Times, Nov. 14, 1971, p. 60, col. 2. He adds: "In Wyoming . . . the Forest Service is very much . . . nursemaid . . . to the lumber industry . . . ." Hearings on Management Practices on the Public Lands before the Subcommittee on Public Lands of the Senate Committee on Interior and Insular Affairs, pt. 1, p. 7 (1971).

Senator Jennings Randolph offers a similar criticism of the leveling by lumber companies of large portions of the Monongahela National Forest in West Virginia. Id., at 9. See also 116 Cong. Rec. 36971 (reprinted speech of Sen. Jennings Randolph concerning Forest Service policy in Monongahela National Forest). To investigate similar controversy surrounding the Service's management of the Bitterroot National Forest in Montana, Senator Lee Metcalf recently asked forestry professionals at the University of Montana to study local harvesting practices. The faculty group concluded that public dissatisfaction had arisen from the Forest Service's "overriding concern for sawtimber production" and its "insensitivity to the related forest uses and to the . . . public's interest in environmental values." S. Doc. No. 91–115, p. 14 (1970). See also Behan, Timber Mining: Accusation or Prospect? American Forests, Nov. 1971, p. 4 (additional comments of faculty participant); Reich, The Public and the Nation's Forests, 50 Calif. L. Rev. 381–400 (1962).

Former Secretary of the Interior Walter Hickel similarly faulted clearcutting as excusable only as a money-saving harvesting practice for large lumber corporations. W. Hickel, Who Owns America? 130 (1971). See also Risser, The U.S. Forest Service: Smokey's Strip Miners, 3 The Washington Monthly, Dec. 1971, p. 16. And at least one Forest Service study team shares some of these criticisms of clearcutting. U.S. Dept. of Agriculture, Forest Management in Wyoming 12 (1971). See also Public Land Law Review Comm'n, Report to the President and to the Congress 44 (1970); Chapman, Effects of Logging upon Fish Resources of the West Coast, 60 J. of Forestry 533 (1962).

A third category of criticism results from the Service's huge backlog of delayed reforestation projects. It is true that Congress has underfunded replanting programs of the Service but it is also true that the Service and lumber companies have regularly ensured that Congress fully funds budgets requested for the Forest Service's "timber sales and management." M. Frome, The Environment and Timber Resources, in What's Ahead for Our Public Lands? 23, 24 (H. Pyles ed. 1970).

8. Permitting a court to appoint a representative of an inanimate object would not be significantly different from customary judicial appointments of guardians ad litem, executors, conservators, receivers, or counsel for indigents.

   The values that ride on decisions such as the present one are often not appreciated even by the so-called experts.

> "A teaspoon of living earth contains 5 million bacteria, 20 million fungi, one million protozoa, and 200,000 algae. No living human can predict what vital miracles may be locked in this dab of life, this stupendous reservoir of genetic materials that have evolved continuously since the dawn of the earth. For example, molds have existed on earth for about 2 billion years. But only in this century did we unlock the secret of the penicillins, tetracyclines, and other antibiotics from the lowly molds, and thus fashion the most powerful and effective medicines ever discovered by man. Medical scientists still wince at the thought that we might have inadvertently wiped out the rhesus monkey, medically, the most important research animal on earth. And who knows what revelations might lie in the cells of the blackback gorilla nesting in his eyrie this moment in the Virunga Mountains of Rwanda? And what might we have learned from the European lion, the first species formally noted (in 80 A.D.) as extinct by the Romans?
>
> "When a species is gone, it is gone forever. Nature's genetic chain, billions of years in the making, is broken for all time." Conserve—Water, Land and Life, Nov. 1971, p. 4.

Aldo Leopold wrote in Round River 147 (1953):

> "In Germany there is a mountain called the Spessart. Its south slope bears the most magnificent oaks in the world. American cabinetmakers, when they want the last word in quality, use Spessart oak. The north slope, which should be the better, bears an indifferent stand of Scotch pine. Why? Both slopes are part of the same state forest; both have been managed with equally scrupulous care for two centuries. Why the difference?
>
> "Kick up the litter under the oaks and you will see that the leaves rot almost as fast as they fall. Under the pines, though, the needles pile up as a thick duff; decay is much slower. Why? Because in the Middle Ages the south slope was preserved as a deer forest by a hunting bishop; the north slope was pastured, plowed, and cut by settlers, just as we do with our woodlots in Wisconsin and Iowa today. Only after this period of abuse was the north slope replanted to pines. During this period of abuse something happened to the microscopic flora and fauna of the soil. The number of species was greatly reduced, i. e., the digestive apparatus of the soil lost some of its parts. Two centuries of conservation have not sufficed to restore these losses. It required the modern microscope, and a

century of research in soil science, to discover the existence of these 'small cogs and wheels' which determine harmony or disharmony between men and land in the Spessart."

9. Senator Cranston has introduced a bill to establish a 35,000-acre Pupfish National Monument to honor the pupfish which are one inch long and are useless to man. S. 2141, 92d Cong., 1st Sess. They are too small to eat and unfit for a home aquarium. But as Michael Frome has said:

   "Still I agree with Senator Cranston that saving the pupfish would symbolize our appreciation of diversity in God's tired old biosphere, the qualities which hold it together and the interaction of life forms. When fishermen rise up united to save the pupfish they can save the world as well." Field & Stream, Dec. 1971, p. 74.

Notes to Page 211

1. Environmental Defense Fund, Inc. v. Hardin, 138 U.S. App. D.C. 391, 394–395, 428 F.2d 1093, 1096–1097 (1970); Citizens Committee for the Hudson Valley v. Volpe, 425 F.2d 97, 101–105 (CA2 1970), cert. denied, 400 U.S. 949 ; Scenic Hudson Preservation Conf. v. FPC, 354 F.2d 608, 615–617 (CA2 1965); Izaak Walton League v. St. Clair, 313 F. Supp. 1312, 1316–1317 (Minn. 1970); Environmental Defense Fund, Inc. v. Corps of Engineers, 324 F. Supp. 878, 879–880 (DC 1971); Environmental Defense Fund, Inc. v. Corps of Engineers, 325 F. Supp. 728, 734–736 (ED Ark. 1970–1971); Sierra Club v. Hardin, 325 F. Supp. 99, 107–112 (Alaska 1971); Upper Pecos Assn. v. Stans, 328 F. Supp. 332, 333–334 (N. Mex. 1971); Cape May County Chapter, Inc., Izaak Walton League v. Macchia, 329 F. Supp. 504, 510–514 (N. J. 1971). See National Automatic Laundry & Cleaning Council v. Shultz, 143 U.S. App. D.C. 274, 278–279, 443 F.2d 689, 693–694 (1971); West Virginia Highlands Conservancy v. Island Creek Coal Co., 441 F.2d 232, 234–235 (CA4 1971); Environmental Defense Fund Inc. v. HEW, 138 U.S. App. D.C. 381, 383 n. 2, 428 F.2d 1083, 1085 n. 2 (1970); Honchok v. Hardin, 326 F. Supp. 988, 991 (Md. 1971).
2. "No man is an Iland, intire of itselfe; every man is a peece of the Continent, a part of the maine; if a Clod bee washed away by the Sea, Europe is the lesse, as well as if a Promontorie were, as well as if a Mannor of thy friends or of thine owne were; any man's death diminishes me, because I am involved in Mankinde; And therefore never send to know for whom the bell tolls; it tolls for thee." Devotions XVII.

25

Edward Abbey (1927–1989)

After graduating from high school in Pennsylvania, Abbey was drafted into the Army and spent nearly two years in occupied Italy before receiving his discharge in 1946. The following year he moved out west and matriculated at the University of New Mexico, where he received a B.A. in Philosophy in 1951. Abbey returned to the University of New Mexico several years later, receiving a master's degree in 1959 (his thesis was on "Anarchism and the Morality of Violence"), but his primary interest was in writing fiction. Abbey's first novel, Jonathan Troy *(1954) was critically and commercially unsuccessful, but a film version of his second novel,* The Brave Cowboy: An Old Tale in a New Time *(1956), retitled* Lonely are the Brave, *was made in 1962. Abbey's literary breakthrough, however, did not come until 1968 when he published his first work of nonfiction,* Desert Solitaire. *That book, centered around Abbey's experience as a seasonal park ranger in Arches National Monument, gained Abbey a reputation as a nature writer. Abbey resisted this categorization, referring to his nonfiction writings as "personal history" rather than nature writing.*

Abbey's politics are also somewhat difficult to categorize given his fondness for ambiguity, paradox, and the outrageously provocative. He was an environmentalist, but his views on topics such as gun control and immigration often alienated him from other environmentalists. In the "ideal realm," said Abbey, he was an anarchist, and in a 1951 journal entry he identified the literary theme that most intrigued him: "The harried anarchist, a wounded wolf, struggling toward the green hills, or the black-white alpine mountains, or the purple-golden desert range and liberty. Will he make it? Or will the FBI shoot him down on the very threshold of wilderness and freedom?" Abbey explored this theme in many of his novels, including Fire on the Mountain *(1962),* The Monkey Wrench Gang *(1975),* Good News *(1980), and the posthumously published* Hayduke Lives! *(1990). In* The Monkey Wrench Gang, *the novel's protagonists travel throughout the four corners region of the southwest engaging in acts of "monkey wrenching" or "ecotage" aimed against the forces of development. The novel inspired the founding of EarthFirst! and similar direct action groups whose tactics have been disavowed by many mainstream environmentalists. It is important to understand that Abbey's brand of environmentalism does not spring only from his antagonism toward the exploitation and development of wilderness areas (he was fond of saying that "growth for the sake of growth is the ideology of the cancer cell") but out of a deep-seated fear of governmental oppression. In essays such as "Freedom and Wilderness, Wilderness and Freedom," Abbey adds this political component—wilderness as a refuge from government tyranny—to his other, more familiar, arguments for wilderness protection.*

"Freedom and Wilderness, Wilderness and Freedom" (1977)

When I lived in Hoboken, just across the lacquered Hudson from Manhattan, we had all the wilderness we needed. There was the waterfront with its decaying piers and abandoned warehouses, the jungle of bars along River Street and Hudson Street, the houseboats, the old ferry slips, the mildew-green cathedral of the Erie-Lackawanna Railway terminal. That was back in 1964–65: then came Urban Renewal, which ruined everything left lovable in Hoboken, New Jersey.

What else was there? I loved the fens, those tawny marshes full of waterbirds, mosquitoes, muskrats, and opossums that intervened among the black basaltic rocks between Jersey City and Newark, and somewhere back of Union City on the way to gay, exotic, sausage-packing, garbage-rich Secaucus. I loved also and finally and absolutely, as a writer must love any vision of eschatological ultimates, the view by twilight from the Pulaski Skyway (Stop for Emergency Repairs Only) of the Seventh Circle of Hell. Those melancholy chemical plants, ancient as acid, sick as cyanide, rising beyond the cattails and tules; the gleam of oily waters in the refineries' red glare; the desolation of the endless, incomprehensible uninhabitable (but inhabited) slums of Harrison, Newark, Elizabeth; the haunting and sinister odors on the wind. Rust and iron and sunflowers in the tangled tracks, the great grimy sunsets beyond the saturated sky . . . . It will all be made, someday, a national park of the mind, a rigid celebration of industrialism's finest frenzy.

We tried north too, up once into the Catskills, once again to the fringe of the Adirondacks. All I saw were Private Property Keep Out This Means You signs. I live in a different country now. Those days of longing, that experiment in exile, are all past. The far-ranging cat returns at last to his natural, native habitat. But what wilderness there was in those bitter days I learned to treasure. Foggy nights in greasy Hoboken alleyways kept my soul alert, healthy and aggressive, on edge with delight.

The other kind of wilderness is also useful. I mean now the hardwood forests of upper Appalachia, the overrated mountains of Colorado, the burnt sienna hills of South Dakota, the raw umber of Kansas, the mysterious swamps of Arkansas, the porphyritic mountains of purple Arizona, the mystic

Reprinted with permission of Dutton, a division of Penguin Group (USA) Inc., "Freedom and Wilderness, Wilderness and Freedom," from *The Journey Home* by Edward Abbey. Copyright © 1977 by Edward Abbey.

desert of my own four-cornered country—this and 347 other good, clean, dangerous places I could name.

Science is not sufficient. "Ecology" is a word I first read in H. G. Wells twenty years ago and I still don't know what it means. Or seriously much care. Nor am I primarily concerned with nature as living museum, the preservation of spontaneous plants and wild animals. The wildest animal I know is you, gentle reader, with this helpless book clutched in your claws. No, there are better reasons for keeping the wild wild, the wilderness open, the trees up and the rivers free, and the canyons uncluttered with dams.

We need wilderness because we are wild animals. Every man needs a place where he can go to go crazy in peace. Every Boy Scout troop deserves a forest to get lost, miserable, and starving in. Even the maddest murderer of the sweetest wife should get a chance for a run to the sanctuary of the hills. If only for the sport of it. For the terror, freedom, and delirium. Because we need brutality and raw adventure, because men and women first learned to love in, under, and all around trees, because we need for every pair of feet and legs about ten leagues of naked nature, crags to leap from, mountains to measure by, deserts to finally die in when the heart fails.

The prisoners in Solzhenitsyn's labor camps looked out on the vast Siberian forests—within those shadowy depths lay the hope of escape, of refuge, of survival, of hope itself—but guns and barbed wire blocked the way. The citizens of our American cities enjoy a high relative degree of political, intellectual, and economic liberty; but if the entire nation is urbanized, industrialized, mechanized, and administered, then our liberties continue only at the sufferance of the technological megamachine that functions both as servant and master, and our freedoms depend on the pleasure of the privileged few who sit at the control consoles of that machine. What makes life in our cities at once still tolerable, exciting, and stimulating is the existence of an alternative option, whether exercised or not, whether even appreciated or not, of a radically different mode of being *out there*, in the forests, on the lakes and rivers, in the deserts, up in the mountains.

Who needs wilderness? Civilization needs wilderness. The idea of wilderness preservation is one of the fruits of civilization, like Bach's music, Tolstoy's novels, scientific medicine, novocaine, space travel, free love, the double martini, the secret ballot, the private home and private property, the public park and public property, freedom of travel, the Bill of Rights, peppermint toothpaste, beaches for nude bathing, the right to own and bear arms, the right not to own and bear arms, and a thousand other good things one could name, some of them trivial, most of them essential, all of them vital to that great, bubbling, disorderly, anarchic, unmanageable diversity of opinion, expression, and ways of living which free men and women love, which is their breath of life, and which the authoritarians of church and state and war and sometimes even art despise and always have despised. And feared.

The permissive society? What else? I love America because it *is* a confused, chaotic mess—and I hope we can keep it this way for at least another thousand years. The permissive society is the free society, the open society.

Who gave us permission to live this way? Nobody did. *We* did. And that's the way it should be—only more so. The best cure for the ills of democracy is more democracy.

The boundary around a wilderness area may well be an artificial, self-imposed, sophisticated construction, but once inside that line you discover the artificiality beginning to drop away; and the deeper you go, the longer you stay, the more interesting things get—sometimes fatally interesting. And that too is what we want: Wilderness is and should be a place where, as in Central Park, New York City, you have a fair chance of being mugged and buggered by a shaggy fellow in a fur coat—one of Pooh Bear's big brothers. To be alive is to take risks; to be always safe and secure is death.

Enough of these banalities—no less true anyhow—which most of us embrace. But before getting into the practical applications of this theme, I want to revive one more argument for the preservation of wilderness, one seldom heard but always present, in my own mind at least, and that is the political argument.

Democracy has always been a rare and fragile institution in human history. Never was it more in danger than now, in the dying decades of this most dangerous of centuries. Within the past few years alone we have seen two more relatively open societies succumb to dictatorship and police rule—Chile and India. In all of Asia there is not a single free country except Israel—which, as the Arabs say, is really a transplanted piece of Europe. In Africa, obviously going the way of Latin America, there are none. Half of Europe stagnates under one-man or one-party domination. Only Western Europe and Britain, Australia and New Zealand, perhaps Japan, and North America can still be called more or less free, open, democratic societies.

As I see it, our own nation is not free from the danger of dictatorship. And I refer to internal as well as external threats to our liberties. As social conflict tends to become more severe in this country—and it will unless we strive for social justice—there will inevitably be a tendency on the part of the authoritarian element—always present in our history—to suppress individual freedoms, to utilize the refined techniques of police surveillance (not excluding torture, of course) in order to preserve—not wilderness!—but the status quo, the privileged positions of those who now so largely control the economic and governmental institutions of the United States.

If this fantasy should become reality—and fantasies becoming realities are the story of the twentieth century—then some of us may need what little wilderness remains as a place of refuge, as a hideout, as a base from which to carry on guerrilla warfare against the totalitarianism of my nightmares. I hope it does not happen; I believe we will prevent it from happening; but if it should, then I, for one, intend to light out at once for the nearest national forest, where I've been hiding cases of peanut butter, home-brew, ammunition, and C-rations for the last ten years. I haven't the slightest doubt that the FBI, the NSA, the CIA, and the local cops have dossiers on me a yard thick. If they didn't, I'd be insulted. Could I survive in the wilderness? I don't know—but I do know I could never survive in prison.

Could we as a people survive without wilderness? To consider that question we might look at the history of modern Europe, and of other places. As the Europeans filled up their small continent, the more lively among them spread out over the entire planet, seeking fortune, empire, a new world, a new chance—but seeking most of all, I believe, for adventure, for the opportunity of self-testing. Those nations that were confined by geography, bottled up, tended to find their outlet for surplus energy through war on their neighbors; the Germans provide the best example of this thesis. Nations with plenty of room for expansion, such as the Russians, tended to be less aggressive toward their neighbors.

In Asia we can see the same human necessities at work in somewhat different forms. Japan might be likened to Germany; a small nation with a large, ever-growing, vigorous, and intelligent population. Confined by the sea, their open spaces long ago occupied and domesticated, the Japanese like the Germans turned to war upon their neighbors, particularly China, Korea, and Oriental Russia; and when that was not enough to fully engage their surplus energies, they became an oceanic power, which soon brought them into conflict with two other oceanic powers—Britain and the United States. Defeated in war, the Japanese turned their undefeated energies into industry and commerce, becoming a world power through trade. But that kind of adventure is satisfactory for only a small part of the population; and when the newly prosperous Japanese middle class becomes bored with tourism, we shall probably see some kind of civil war or revolution in Japan—perhaps within the next twenty years.

Something of that sort may be said to have already happened in China. Powerless to wage war upon their neighbors, the Chinese waged war upon themselves, class against class, the result a triumphant revolution and the construction of a human society that may well become, unfortunately, the working model for all. I mean the thoroughly organized society, where all individual freedom is submerged to the needs of the social organism.

The global village and the technological termitorium. More nightmares! I do not believe that human beings would or could long tolerate such a world. The human animal is almost infinitely adaptable—but there must be limits to our adaptability, limits beyond which, if we can survive them at all, we would survive only by sacrificing those qualities that distinguish the human from that possible cousin of the future: the two-legged, flesh-skinned robot, his head, her head, its head wired by telepathic radio to a universal central control system.

One more example: What happened to India when its space was filled, its wilderness destroyed? Something curiously different from events in Europe, China, or Japan; unable to expand outward in physical space, unable or unwilling (so far) to seek solutions through civil war and revolution, the genius of India—its most subtle and sensitive minds—sought escape from unbearable reality by rocket flights of thought into the inner space of the soul, into a mysticism so deep and profound that a whole nation, a whole people, have been paralyzed for a thousand years by awe and adoration.

Now we see something similar happening in our own country. A tiny minority, the technological elite, blast off for the moon, continuing the traditional European drive for the conquest of physical space. But a far greater number, lacking the privileges and luck and abilities of the Glenns and the Armstrongs and their comrades, have attempted to imitate the way of India: When reality becomes intolerable, when the fantasies of nightmare become everyday experience, then deny that reality, obliterate it, and escape, escape, escape, through drugs, through trance and enchantment, through magic and madness, or through study and discipline. By whatever means, in some cases by *any* means, escape this crazy, unbearable, absurd playpen of the senses—this gross 3-D, grade-B, X-rated, porno flick thrust upon us by CBS News, *Time, Newsweek*, the *New York Times, Rolling Stone*, and the *Sierra Club Bulletin*—seeking refuge in a nicer universe just next door, around some corner of the mind and nervous system, deep in the coolest cells of the brain. If all is illusion then nothing matters, or matters much; and if nothing matters then peace, of a sort, is possible, striving becomes foolish, and we can finally relax, at last, into that bliss which passeth understanding, content as pigs on a warm manure pile. Until the man comes with the knife, to carry the analogy to its conclusion, until pig-sticking time rolls around again and the fires are lit under the scalding tubs.

You begin to see the outline of my obsessions. Every train of thought seems to lead to some concentration camp of nightmare. But I believe there are alternatives to the world of nightmare. I believe that there are better ways to live than the traditional European-American drive for power, conquest, domination; better ways than the horrifying busyness of the Japanese; better ways than the totalitarian communes of the Chinese; better ways than the passive pipe dreams of Hindu India, that sickliest of all nations.

I believe we can find models for a better way both in the past and the present. Imperfect models, to be sure, each with its grievous faults, but better all the same than most of what passes for necessity in the modern world. I allude to the independent city-states of classical Greece; to the free cities of medieval Europe; to the small towns of eighteenth and nineteenth-century America; to the tribal life of the American Plains Indians; to the ancient Chinese villages recalled by Lao-tse in his book, *The Way*.

I believe it is possible to find and live a balanced way of life somewhere halfway between all-out industrialism on the one hand and a make-believe pastoral idyll on the other. I believe it possible to live an intelligent life in our cities—if we make them fit to live in—if we stop this trend toward joining city unto city until half the nation and half the planet becomes one smog-shrouded, desperate and sweating, insane and explosive urbanized concentration camp.

According to my basic thesis, if it's sound, we can avoid the disasters of war, the nightmare of the police state and totalitarianism, the drive to expand and conquer, if we return to this middle way and learn to live for a while, say at least a thousand years or so, just for the hell of it, just for the fun of it, in some sort of steady-state economy, some sort of free, democratic, wide-open society.

As we return to a happier equilibrium between industrialism and a rural-agrarian way of life, we will of course also encourage a gradual reduction of the human population of these states to something closer to the optimum: perhaps half the present number. This would be accomplished by humane social policies, naturally, by economic and taxation incentives encouraging birth control, the single-child family, the unmarried state, the community family. Much preferable to war, disease, revolution, nuclear poisoning, etc., as population control devices.

What has all this fantasizing to do with wilderness and freedom? We can have wilderness without freedom; we can have wilderness without human life at all; but we cannot have freedom without wilderness, we cannot have freedom without leagues of open space beyond the cities, where boys and girls, men and women, can live at least part of their lives under no control but their own desires and abilities, free from any and all direct administration by their fellow men. "A world without wilderness is a cage," as Dave Brower says.

I see the preservation of wilderness as one sector of the front in the war against the encroaching industrial state. Every square mile of range and desert saved from the strip miners, every river saved from the dam builders, every forest saved from the loggers, every swamp saved from the land speculators means another square mile saved for the play of human freedom.

All this may seem utopian, impossibly idealistic. No matter. There comes a point at every crisis in human affairs when the ideal must become the real—or nothing. It is my contention that if we wish to save what is good in our lives and give our children a taste of a good life, we must bring a halt to the ever-expanding economy and put the growth maniacs under medical care.

Let me tell you a story.

A couple of years ago I had a job. I worked for an outfit called Defenders of Fur Bearers (now known as Defenders of Wildlife). I was caretaker and head janitor of a 70,000-acre wildlife refuge in the vicinity of Aravaipa Canyon in southern Arizona. The Whittell Wildlife Preserve, as we called it, was a refuge for mountain lion, javelina, a few black bear, maybe a wolf or two, a herd of whitetail deer, and me, to name the principal fur bearers.

I was walking along Aravaipa Creek one afternoon when I noticed fresh mountain lion tracks leading ahead of me. Big tracks, the biggest lion tracks I've seen anywhere. Now I've lived most of my life in the Southwest, but I am sorry to admit that I had never seen a mountain lion in the wild. Naturally I was eager to get a glimpse of this one.

It was getting late in the day, the sun already down beyond the canyon wall, so I hurried along, hoping I might catch up to the lion and get one good look at him before I had to turn back and head home. But no matter how fast I walked and then jogged along, I couldn't seem to get any closer; those big tracks kept leading ahead of me, looking not five minutes old, but always disappearing around the next turn in the canyon.

Twilight settled in, visibility getting poor. I realized I'd have to call it quits. I stopped for a while, staring upstream into the gloom of the canyon. I could see the buzzards settling down for the evening in their favorite dead cottonwood. I heard the poor-wills and the spotted toads beginning to sing, but of that mountain lion I could neither hear nor see any living trace.

I turned around and started home. I'd walked maybe a mile when I thought I heard something odd behind me. I stopped and looked back—nothing; nothing but the canyon, the running water, the trees, the rocks, the willow thickets. I went on and soon I heard that noise again—the sound of footsteps.

I stopped. The noise stopped. Feeling a bit uncomfortable now—it was getting dark—with all the ancient superstitions of the night starting to crawl from the crannies of my soul, I looked back again.

And this time I saw him. About fifty yards behind me, poised on a sand bar, one front paw still lifted and waiting, stood this big cat, looking straight at me. I could see the gleam of the twilight in his eyes. I was startled as always by how small a cougar's head seems but how long and lean and powerful the body really is. To me, at that moment, he looked like the biggest cat in the world. He looked dangerous. Now I know very well that mountain lions are supposed almost never to attack human beings. I knew there was nothing to fear—but I couldn't help thinking maybe this lion is different from the others. Maybe he knows we're in a wildlife preserve, where lions can get away with anything. I was not unarmed; I had my Swiss army knife in my pocket with the built-in can opener, the corkscrew, the two-inch folding blade, the screwdriver. Rationally there was nothing to fear; all the same I felt fear.

And something else too: I felt what I always feel when I meet a large animal face to face in the wild: I felt a kind of affection and the crazy desire to communicate, to make some kind of emotional, even physical contact with the animal. After we'd stared at each other for maybe five seconds—it seemed at the time like five minutes—I held out one hand and took a step toward the big cat and said something ridiculous like, "Here, kitty, kitty." The cat paused there on three legs, one paw up as if he wanted to shake hands. But he didn't respond to my advance.

I took a second step toward the lion. Again the lion remained still, not moving a muscle, not blinking an eye. And I stopped and thought again and this time I understood that however the big cat might secretly feel, I myself was not yet quite ready to shake hands with a mountain lion. Maybe someday. But not yet. I retreated.

I turned and walked homeward again, pausing every few steps to look back over my shoulder. The cat had lowered his front paw but did not follow me. The last I saw of him, from the next bend of the canyon, he was still in the same place, watching me go. I hurried on through the evening, stopping now and then to look and listen, but if that cat followed me any further I could detect no sight or sound of it.

I haven't seen a mountain lion since that evening, but the experience remains shining in my memory. I want my children to have the opportunity for that kind of experience. I want my friends to have it. I want even our enemies to have it—they need it most. And someday, possibly, one of our children's children will discover how to get close enough to that mountain lion to shake paws with it, to embrace and caress it, maybe even teach it something, and to learn what the lion has to teach us.

26

Luella N. Kenny (1937–)

In the early 1900s, William T. Love devised a plan to dig a short canal between the upper and lower Niagara Rivers (near the city of Niagara Fall, New York) to generate inexpensive power for what he hoped would be a model community. The project was soon abandoned, however, and in the 1920s the canal was used as a dumpsite for industrial chemicals that had been stored in metal drums. In 1953 the Hooker Chemical Company (later purchased by Occidental Petroleum) closed up Love Canal and sold it to the city of Niagara Falls for one dollar. Within a few years many homes and an elementary school had been built on the site of the dump. By the 1970s, residents noticed heavy chemical odors and groundwater contamination in the community as well as an extraordinary number of birth defects, miscarriages, and cancer. Following an investigation by the New York State Department of Health it was determined that the area was heavily contaminated with numerous chemical compounds including several known carcinogens. On August 2, 1978 New York State declared a state of emergency and Governor Hugh Carey ordered that the homes closest to the canal be evacuated.

To a remarkable extent the protest over Love Canal was driven largely by the residents of the community themselves, and their example has given rise to numerous grassroots organizations throughout the country that have been formed in response to local environmental issues. Love Canal Activists such as Lois Gibbs, Luella Kenny, Dr. Beverly Paigen, and Sister Margeen Hoffman not only helped to spearhead the attempt of the residents to spur the government into action, but also worked to make sure that the company responsible for the disaster be held responsible for its actions. The disaster at Love Canal was also a key factor in the federal government's creation of the Superfund for the cleanup of contaminated sites and that law's provision requiring the party responsible for the polluted site to bear the cost of its remediation.

Statement to the Annual Meeting of Occidental Petroleum Share Holders: Corporate Responsibility Resolution May 21, 1980

My name is Luella Kenny. I am a cancer research assistant at Roswell Park Memorial Institute in Buffalo, New York. From July 1st 1969 to September 5, 1979 I resided at 1064 96th Street, Niagara Falls, New York, which is located approximately 0.1 of a mile from the northern boundary of the Love Canal. My husband and I with our two surviving sons were forced to abandon this residence because of the presence of toxins that had migrated from the Love Canal. Since that time we have lived a vagabond existence waiting for this problem to be resolved. An old streambed, which intersected with Love Canal, runs through our property. This streambed is now filled and is part of our yard. In addition, at the back edge of our property is Black Creek, which has been found to be contaminated with chemicals by the Environmental Protection Agency and by the New York State Health Department. Also located on this property is a storm sewer which drains the area north of the Love Canal. Large amounts of Dioxin were found where this storm sewer empties into Black Creek. My sons spent many hours playing in the creek by this storm sewer. Our seven year old son died October 4th, 1978 from complications that resulted from nephrosis. This spring E.P.A. erected a six foot fence in our yard and along Black Creek because of the hazard.

Jon became ill on June 6th 1978. Initially his illness was diagnosed as an allergy, however on July 1st, 1980 he was diagnosed as having nephrosis. Nephrosis in its early stages is often masked by symptoms resembling allergies. John responded to treatment for nephrosis and was referred to Children's Hospital in Buffalo, New York because of the chronic nature of the disease. Jon had three relapses in two and a half months and then he developed convulsions, visual hallucinations, and eventually a massive pulmonary embolism. Jon's death was caused by a cardiac arrest brought on by the exertion of trying to breathe.

At the time of Jon's death we had no idea that it could be linked to chemical toxicity. We requested an autopsy because we wanted to know why our son had died when we had been told all along that nephrosis was nothing to worry about. To quote the urologist at Children's Hospital, "nephrosis is the best disease a child can have, because it can be cured." However, this same doctor was puzzled by the fact that Jon's symptoms were not typical, particularly the convulsions, nor did he respond to treatment as expected.

After Jon's death we read in the newspaper that the state of New York was going to investigate his death. It was at this time that my husband and I began to try to learn more about the disease. We began delving into medical journals, and corresponding with leading research groups in the field of nephrosis. We were shocked to find that during the past ten years there have been countless reports of people developing nephrosis when they were exposed to chemicals. We also did some research into dioxin toxicity and discovered that many of Jon's autopsy findings were related to dioxin poisoning; for example a shrunken thymus.

Since we left our home in September our two sons have shown a remarkable improvement. The older boy who had frequent headaches is much better, and our younger son's appetite has returned.

Dr. Beverly Paigen from Roswell Park Memorial Institute compiled an epidemeological study of the Love Canal Area. This study clearly shows the higher incidences of miscarriages, birth defects, nervous disorders and suicides just to name a few of the many illnesses that are being experienced by the residents. Many of these illnesses follow the filled in streambeds indicating a migration of chemicals. These illnesses were even more graphically portrayed when we were living in hotels last fall. It's one thing, to read the statistics, but it is something else when you have hundreds of people together under a common roof with their related illnesses.

It's amazing all of these illnesses are prefaced with the term ideopathic, that is, origin unknown. Jon's illness is referred to in medical journals as ideopathic nephrosis. Malissa Gibbs, the four year old daughter of the Love Canal Homeowners Association president, was rushed to the hospital last month when her blood platelet count dropped from a normal of 150,000–400,000 to 1000. She was diagnosed as having a blood disease called I.T.P. The I of course stands for idiopathic.

The following words are truer today than what they were in 1776 when Thomas Paine penned them in his American Crisis #1, he wrote "These are the times to try men's souls." Inflation has forced both individual consumers and giant corporations to reassess their "style of living." The current political unrests that exist all over the world have left us in fear of another world war.

However, we won't have to worry about the luxuries we can't afford because of inflation, and why worry about an enemy who will destroy us when we are self-destructing. We don't need sophisticated nuclear weapons; all we need are the multitude of dumps strategically placed all over the country that will insidiously destroy everything and everyone in its path.

27

Dolores LaChapelle (1925–)

Dolores LaChapelle grew up in Denver, Colorado, and graduated Phi Beta Kappa from Denver University in 1947. In 1975 LaChapelle founded the Way of the Mountain Learning Center in Silverton, Colorado, where she teaches Tai Chi, writes, climbs mountains, skis, and leads experiential ecology workshops. She has become widely known as one of the first American woman to advocate Deep Ecology, a subject that she has written about widely in books such as Earth Festivals *(1976) and* Earth Wisdom *(1978). In many of her works, LaChapelle makes the case that it will take more than legislation and improved technologies to reestablish a sustainable relationship with the land, but will require a personal connection fostered through ritual and ceremony, as was the case throughout much of preindustrial history.*

"Ritual is Essential: Seeing Ritual and Ceremony as Sophisticated Social and Spiritual Technology" (1984)

> But there's more than just solving the how-to problems. I've often said that if we're going to have a real rural renaissance I'd just take the how-to problems for granted. The first thing I'd provide would be festivals.
>
> Ralph Borsodi

> I will say leave a good quarter of the time for feast and celebration or your soul will die.
>
> Francois Monnet,
> IN CONTEXT, No. 1

Most native societies around the world had three common characteristics: they had an intimate, conscious relationship with their place; they were stable

Reprinted with permission of the author and *In Context: A Quarterly of Humane Sustainable Culture* (Spring 1984), 39. Copyright © 1984, 1997 by Context Institute.

"sustainable" cultures, often lasting for thousands of years; and they had a rich ceremonial and ritual life. They saw these as intimately connected. Out of the hundreds of examples of this, consider the following:

- The Tukano Indians of the Northwest Amazon River basin, guided by their shamans who are conscious ecologists, make use of various myths and rituals that prevent over-hunting or over-fishing. They view their universe as a circuit of energy in which the entire cosmos participates. This basic circuit of energy consists of "a limited quantity of procreative energy that flows continually between man and animals, between society and nature." Reichel-Dolmatoff, the Columbian anthropologist, notes that the Tukano have very little interest in exploiting natural resources more effectively but are greatly interested in "accumulating more factual knowledge about biological reality and, above all, about knowing what the physical world requires from men."[1]
- The !Kung people of the Kalahari Desert have been living in exactly the same place for 11,000 years! They have very few material belongings but their ritual life is one of the most sophisticated of any group.[2]
- Roy Rappaport has shown that the rituals of the Tsembaga of New Guinea allocate scarce protein for the humans who need it without causing irreversible damage to the land.[3]
- The longest-inhabited place in the United States is the Hopi village of Oraibi. At certain times of the year they may spend up to half their time in ritual activity.[4]
- Upon the death of their old *cacique*, Santa Ana Pueblo in New Mexico recently elected a young man to take over as the new *cacique*. For the rest of his life he will do nothing else whatsoever but take care of the ritual life of the pueblo. All his personal needs will be taken care of by the tribe. But he cannot travel any further than sixty miles or one hour distance—his presence is that important to the ongoing life of the Pueblo. They know that it is ritual which embodies the people.[5]

Our Western European industrial culture provides a striking contrast to all these examples. We have idolized ideals, rationality, and a limited kind of "practicality," and have regarded the conscious rituals of these other cultures as at best frivolous curiosities. The results are all too evident. We've been here only a few hundred years and we have already done irreparable damage to vast areas of the country now called the U. S. As Gregory Bateson notes, "mere purposive rationality is necessarily pathogenic and destructive of life."[6]

We have tried to relate to the world around us through only the left side of our brain, and we are clearly failing. It we are to re-establish a viable relationship, we will need to rediscover the wisdom of these other cultures who knew that their relationship to the land and to the natural world required the whole of their being. What we call their "ritual and ceremony" was a sophisticated social and spiritual technology, refined through many thousands of years of experience, that maintained this relationship much more successfully than we are.

The human race has forgotten so much in the last 200 years that we hardly know where to begin. But it helps to begin remembering. In the first place, *all* traditional cultures, even our own long-ago Western European cultural ancestors, had seasonal festivals and rituals.

The true origin of most of our modern major holidays dates back to these seasonal festivals. There are four major festivals: winter and summer solstice (when the sun reverses its travels) and spring and autumn equinox (when night and day are equal). But in between each of these major holidays are the "cross quarter days." For example, spring equinox comes around March 21 or 22 but spring is only barely beginning at that time in Europe. True spring (reliably warm spring) doesn't come until later. This is the cross quarter day—May 1—which Europe celebrated with maypoles, gathering flowers, and fertility rites. May became the month of Mary after the Christian church took over and May crownings and processions were devoted to Mary instead of the old "earth goddesses." Summer solstice comes on June 21. The next cross quarter day is Lammas Day in early August. This is the only festival that our country does not celebrate in any way. The Church put the Feast of the Assumption on this day to honor Mary. Fall equinox comes on Sept. 21—the cross quarter day is Hallowe'en, the ancient Samhain of the Celts. Then comes winter solstice—the sun's turnaround point from darkness to light. The cross quarter day between the solstice and spring equinox is in early February—now celebrated in the Church as Candlemas.[7]

The purpose of seasonal festivals is periodically to revive the *topocosm*. Gaster coined this word from the Greek—*topo* for place and *cosmos* for world order. Topocosm thus means "the world order of a particular place." The topocosm is the entire complex of any given locality conceived as a living organism—not just the human community but the total community—the plants, animals, and soils of the place. The topocosm is not only the actual and present living community but also that continuous entity of which the present community is but the current manifestation.[8]

Seasonal festivals make use of myths, art, dance, and games. All of these aspects of ritual serve to connect—to keep open the essential connections within ourselves. Festivals connect the conscious with the unconscious, the right and left hemispheres of the brain, the cortex with the older three brains (this includes the Oriental *tan tien*—four fingers below the navel), as well as connecting the human with the nonhuman: the earth, the sky, the animals and plants.[9]

The next step after seasonal rituals is to acknowledge the non-human co-inhabitants of your place. You can begin by looking into the records of the tribes of Indians who lived there and see what their totem was. Look into the accounts of the early explorers and very early settlers. Barry Lopez relates that the Eskimos told him that their totem animal was always the one who could teach them something they needed to learn.[10]

Salmon is the totem animal for the North Pacific Rim. "Only Salmon, as a species, informs us humans, as a species, of the vastness and unity of the North Pacific Ocean and its rim. . . . Totemism is a method of perceiving power, goodness and mutuality in locale through the recognition of and respect for the vitality, spirit, and interdependence of other species," as Linn House explains. For at least 20,000 years, the Yurok, Chinook, Salish, Kwakiutl, Haida, and Aleut on this side of the rim, and, on the other rim of the Pacific, the Ainu (the primitives of Japan) ordered their daily lives according to the timing of the Salmon population.[11]

Several years ago I did some in-depth study of Celtic myth and discovered that Salmon was the totem animal for the Celts, too. According to their myth, there was a sacred well situated under the sea where the sacred Salmon acquired their supernatural wisdom. The famous Celtic hero, Finn, traditionally obtained his wisdom when he sucked on the thumb he had just burnt when picking up the Salmon he cooked.[12] It is not surprising that Salmon links all these areas. The North Pacific Rim and the British Isles are maritime climates in the northern half of the earth. Here is the perfect way to ritualize the link between planetary villagers around the earth—through their totem animal.

How can we learn from Salmon? One specific way is to reclaim our waterways so that Salmon can again flourish. If we reclaim the water so that Salmon can flourish, we have reclaimed the soil, the plants, and the other species of the ecosystem—restored them to aboriginal health. And by doing this, we would be restoring full health to our children as well.

Linn House feels that the people who live in or near the spawning grounds of Salmon should form associations—not law enforcement agencies such as the State Fish and Game Department—but educational groups and providers of ritual and ceremony that would celebrate the interdependence of species. Linn was a Salmon fisherman on Guemes Island in Washington; he now lives in Northern California where he is restocking Salmon rivers.

What relevance do these kinds of rituals have for people who live in the city? All of us need seasonal and nature rituals wherever we live, but let me give you a specifically urban example.

Siena, Italy, with a population of about fifty-nine thousand people, has the lowest crime rate of any Western city of comparable size. Delinquency, drug addiction, and violence are virtually unknown. Class is not pitted against class nor young against old.

Why? Because it is a tribal, ritualized city organized around the *contrada* (clans)—With names such as Chiocciola (the Snail), Tartule (the Turtle) and the like—and the *Palio* (the annual horse race). The *contrada* function as independent city states. Each has its own flag, its own territorial boundaries, its own discrete identity, church songs, patron saint and rituals. Particular topographical features of each *contrada's* area are ritualized and mythologized. The ritualized city customs extend clear back to the worship of Diana, the Roman goddess of the moon. Her attributes were taken over by the worship of Mary when Christianity came in.

Many famous writers, including Henry James, Ezra Pound, and Aldous Huxley, sensed the energy of the city and its events and tried to write about it. But none of them even faintly grasped the yearlong ritualized life behind it. About one week before the day of the *Palio* race, workers from the city of Siena begin to bring yellow earth (*la terra* from the fields outside Siena) and spread it over the great central square (the Campo) thus linking the city with its origins in the earth of its *place*. In fact, anytime during the course of the year when someone needs to be cheered up, the sad person is told not to worry because soon there will be "la terra in piazza" (soon there will be earth in the square).

The horse race serves two main purposes. In the intense rivalry surrounding the race, each c*ontrada* "rekindles its own sense of identity." The *Palio* also provide the Sienese with an outlet for their aggressions and, as such, is a ritual war. The horse race grew out of games which were actually mimic battles and were used to mark the ends of religious festivals.

The *Palio* is truly a religious event. On this one day of the year the *contrada's* horse is brought into the church of its patron saint. In the act of blessing the horse, the *contrada* is itself blessed. This horse race is the community's greatest rite. "In the *Palio*, all the flames of Hell are transformed into the lights of Paradise," according to a local priest, Don Vittorio.[13]

If we want to build a sustainable culture, it is not enough merely to "go back to the land." That's exactly where our pioneering ancestors lived; as the famous Western painter Charles Russell said, "A pioneer is a man who comes to virgin country, traps off all the fur, kills off the wild meat, plows the roots up. . . . A pioneer destroys things and calls it civilization."

If we are truly to re-connect with the land, we need to change our perceptions and approach more than our location. As long as we limit ourselves to narrow rationality and its limited sense of "practicality," we will be disconnected from the "deep ecology" of our place. As Heidegger explains: "Dwelling is not primarily inhabiting but taking care of and creating that space within which something comes into its own and flourishes."[14] It takes both time and ritual for real dwelling. Likewise, as Roy Rappaport observes, "Knowledge will never replace respect in man's dealings with ecological systems, for the ecological systems in which man participates are likely to be so complex that he may never have sufficient comprehension of their content and structure to permit him to predict the outcome of many of his own acts."[15] Ritual is the focused way in which we both experience and express that respect.

Ritual is essential because it is truly the pattern that connects. It provides communication at all levels—among all the systems within the individual human organism; between people within groups; between one group and another in a city; and throughout all these levels between the human and the nonhuman in the natural environment. Ritual provides us with a tool for learning to think logically, analogically, and ecologically as we move toward a sustainable culture. Most important of all, perhaps, is that during rituals we have the experience, unique in our culture, of neither *opposing* nature or *trying* to be in communion with nature; but of *finding* ourselves within nature, and that is the key to sustainable culture.

NOTES

1. G. Reichel-Dolmatoff, "Cosmology as Ecological Analysis: A View from the Rain Forest," in *Man: Journal of the Royal Anthropological Institute* 2, no. 3 (September 1978).
2. Richard B. Lee, "What Hunters Do for a Living, or, How to Make out on Scarce Resources," in *Man the Hunter*, edited by Richard B. Lee and Irven DeVore (Chicago: Aldine Publishing Co., 1968), 30–43.
3. Roy Rappaport, *Pigs for the Ancestors* (New Haven: Yale University Press, 1968).
4. John J. Collins, *Primitive Religion* (New York: Rowman and Littlefield, 1978).
5. Elizabeth Cogburn, from a personal communication.
6. Gregory Bateson, *Steps to an Ecology of Mind* (New York: Ballantine Books, 1972), 146.
7. Dolores LaChapelle, *Sacred Land Sacred Sex, Rapture of the Deep: Concerning Deep Ecology and Celebrating Life* (Durango, Col.: Kivaki Press, 1992), 231–238.
8. Gaster, Theodore H., *Thespis: Ritual, Myth, and Drama in the Ancient Near East* (New York: W. W. Norton, 1961).
9. Eugene C. d'Aquili and Charles D. Laughlin, Jr., "The Neurobiology of Myth and Ritual," in *The Spectrum of Ritual: A Biogenetic Structural Analysis*, edited by Eugene G. d'Aquili, Charles D. Laughlin, Jr., and John McManus (New York: Columbia University Press, 1979), 152–182.
10. Barry Lopez, from a personal communication.
11. Freeman House, "Totem Salmon," in *North Pacific Rim Alive* (San Francisco: Planet Drum, 1979).
12. Tom Jay, "The Salmon of the Heart," in *Working the Woods, Working the Sea*, edited by Finn Wilcox and Jeremiah Gorsline (Port Townsend, Washington D.C.: 1986), 100–124.
13. A. Dundes and A. Falassi, *La Terra in Piazza: An Interpretation of the Palio of Siena* (Berkeley, Calif.: University of California Press, 1975).
14. Martin Heidegger, *The Piety of Thinking*, translated by James G. Hart and John Maraldo (Bloomington: Indian University Press, 1976), 131.
15. Roy Rappaport, *Ecology, Meaning, and Religion* (Berkeley, Calif.: North Atlantic Books, 1979), 100.

28

Arne Naess (1912–)

Norwegian philosopher Arne Naess is the founder of the Deep Ecology movement, which he describes as "a process of reflection leading to action." Naess was a philosophy professor at the University of Oslo from 1939 to 1969, then went on to a second career as an environmental activist. In contrast to what he calls the "shallow ecology" movement, which concentrates on fighting pollution and resource depletion primarily for the benefit of people in the developed countries, Naess offers a set of principles that emphasizes biospherical egalitarianism, diversity, classlessness, decentralization, and personal commitment and action. Naess's eco-philosophy has gained a small but growing number of proponents in the United States, most notably writers such as George Sessions, William Devall, and Dolores LaChapelle.

"The Deep Ecological Movement: Some Philosophical Aspects" (1986)

1. Deep Ecology on the Defensive

Increasing pressure for continued growth and development have placed the vast majority of environmental professionals on the defensive. By way of illustration:

The field ecologist Ivar Mysterud, who both professionally and vigorously advocated deep ecological principles in the late 1960s, encountered considerable resistance. Colleagues at his university said he should keep to his science and not meddle in philosophical and political matters. He should resist the temptation to become a prominent "popularizer" through mass media exposure. Nevertheless, he persisted and influenced thousands of people (including myself).

Mysterud became a well-known professional "expert" at assessing the damage done when bears killed or maimed sheep and other domestic animals

Reprinted with permission from *Philosophical Inquiry*, Vol. 8, Nos. 1–2 (1986). Copyright © 1986, D.Z. Andripoulis.

in Norway. According to the law, their owners are paid damages. And licensed hunters receive permission to shoot bears if their misdeeds become considerable.[1] Continued growth and development required that the sheep industry consolidate, and sheepowners became fewer, richer, and tended to live in cities. As a result of wage increases, they could not afford to hire shepherds to watch the flocks, so the sheep were left on their own even more than before. Continued growth also required moving sheep to what was traditionally considered "bear territory." In spite of this invasion, bear populations grew and troubles multiplied.

How did Mysterud react to these new problems? Did he set limits to the amount of human/sheep encroachment on bear territory? Did he attempt a direct application of his deep ecological perspective to these issues? Quite the contrary. He adopted what appeared to be a shallow wildlife management perspective, and defended the sheepowners: more money to compensate for losses, quicker compensation, and the immediate hiring of hunters who killed mostly "juvenile delinquent" bears accused of killing many sheep.

Protectors of big carnivores noted with concern the change of Mysterud's public "image"; had he really abandoned his former value priorities? Privately he insisted that he hadn't. But in public he tended to remain silent.

The reason for M.'s unexpected actions was not difficult to find: the force of economic growth was so strong that the laws protecting bears would be changed in a highly unfavorable direction if the sheepowners were not soon pacified by accepting some of their not unreasonable demands. After all, it did cost a lot of money to hire and equip people to locate a flock of sheep which had been harassed by a bear and, further, to prove the bear's guilt. And the bureaucratic procedures involved were time-consuming. M. had not changed his basic value priorities at all. Rather, he had adopted a purely defensive compromise. He stopped promoting his deep ecology philosophy publicly in order to retain credibility and standing among opponents of his principles and to retain his friendships with sheepowners.

And what is true of Mysterud is also true of thousands of other professional ecologists and environmentalists. These people often hold responsible positions in society where they might strengthen responsible environmental policy, but, given the exponential forces of growth, their publications, if any, are limited to narrowly professional and specialized concerns. Their writings are surely competent, but lack a deeper and more comprehensive perspective (although I admit that there are some brilliant exceptions to this).

If professional ecologists persist in voicing their value priorities, their jobs are often in danger, or they tend to lose influence and status among those who are in charge of overall policies.[2] Privately, they admit the necessity for deep and far-ranging changes, but they no longer speak out in public. As a result, people deeply concerned about ecology and the environment feel abandoned and even betrayed by the "experts" who work within the "establishment."

In ecological debates, many participants know a lot about particular conservation policies in particular places, and many others have strong views concerning fundamental philosophical questions of environmental ethics, but only a few have both qualities. When these people are silent, the loss is formidable.

For example, the complicated question concerning how industrial societies can increase energy production with the least undesirable consequences is largely a waste of time if this increase is pointless in relation to ultimate human ends. Thousands of experts hired by the government and other big institutions devote their time to this complicated problem, yet it is difficult for the public to find out or realize that many of these same experts consider the problem to be pointless and irrelevant. What these experts consider relevant are the problems of how to stabilize and eventually decrease consumption without losing genuine quality of life for humans. But they continue to work on the irrelevant problems assigned to them while, at the same time, failing to speak out, because the ultimate power is not in their hands.

2. A Call to Speak Out

What I am arguing for is this: even those who completely subsume ecological policies under the narrow ends of human health and well-being cannot attain their modest aims, at least not fully, without being joined by the supporters of deep ecology. They need what these people have to contribute, and this will work in their favor more often than it will work against them. Those in charge of environmental policies, even if they are resource-oriented (and growth tolerating?) decision makers, will increasingly welcome, if only for tactical and not fundamental reasons, what deep ecology supporters have to say. Even though the more radical ethic may seem nonsensical or untenable to them, they know that its advocates are, in practice, doing conservation work that sooner or later must be done. They concur with the practice even though they operate from diverging theories. The time is ripe for professional ecologists to break their silence and express their deepest concerns more freely. A bolder advocacy of deep ecological concerns by those working within the shallow, resource-oriented environmental sphere is the best strategy for regaining some of the strength of this movement among the general public, thereby contributing, however modestly, to a turning of the tide.

What do I mean by saying that even the more modest aims of shallow environmentalism have a need for deep ecology? We can see this by considering the World Conservation Strategy—prepared by the International Union for the Conservation of Nature and Natural Resources (IUCN) in cooperation with the United Nations Environmental Programme (UNEP) and the World Wildlife Fund (WWF). The argument in this important document is thoroughly anthropocentric in the sense that all its recommendations are justified exclusively in terms of their effects upon human health and basic well-being.[3]

A more ecocentric environmental ethic is also recommended apparently for tactical reasons: "A new ethic, embracing plants and animals as well as people, is required for human societies to live in harmony with the natural world on which they depend for survival and well-being." But such an ethic would surely be more effective if it were acted upon by people who believe in its validity, rather than merely its usefulness. This, I think, will come to be understood more and more by those in charge of educational policies. Quite simply, it is indecent for a teacher to proclaim an ethic for tactical reasons only.

Furthermore, this point applies to all aspects of a world conservation strategy. Conservation strategies are more eagerly implemented by people who love what they are conserving, and who are convinced that what they love is intrinsically lovable. Such lovers will not want to hide their attitudes and values, rather they will increasingly give voice to them in public. They possess a genuine ethics of conservation, not merely a tactically useful instrument for human survival.

In short, environmental education campaigns can fortunately combine human-centered arguments with a practical environmental ethic based on either a deeper and more fundamental philosophic or religious perspective, and on a set of norms resting on intrinsic values. But the inherent strength of this overall position will be lost if those who work professionally on environmental problems do not freely give testimony to fundamental norms.

The above is hortatory in the positive etymological sense of that word. I seek "to urge, incite, instigate, encourage, cheer" (Latin: *hortari*). This may seem unacademic but I consider it justifiable because of an intimate relationship between hortatory sentences and basic philosophical views which I formulate in section 8. To trace what follows from fundamental norms and hypotheses is eminently philosophical.

3. What is Deep Ecology?

The phrase "deep ecology movement" has been used up to this point without trying to define it. One should not expect too much from definitions of movements; think, for example, of terms like "conservatism," "liberalism," or the "feminist movement." And there is no reason why supporters of movements should adhere exactly to the same definition, or to any definition, for that matter. It is the same with characterizations, criteria, or a set of proposed necessary conditions for application of the term or phrase. In what follows, a platform or key terms and phrases, agreed upon by George Sessions and myself, are tentatively proposed as basic to deep ecology.[4] More accurately, the sentences have a double function. They are meant to express important points which the great majority of supporters accept, implicitly or explicitly, at a high level of generality. Furthermore, they express a proposal to the effect that those who solidly reject one or more of these points should not be viewed as supporters of deep ecology. This might result because they are supporters of a shallow (or reform) environmental movement or rather they may simply dislike one or more of the eight points for semantical or other reasons. But they may well accept a different set of points which, to me, has roughly the same meaning, in which case I shall call them supporters of the deep ecology movement, but add that they *think* they disagree (maybe

Henryk Skolimowski is an example of the latter). The eight points are:

1. The well-being and flourishing of human and non-human life on Earth have value in themselves (synonyms: intrinsic value, inherent worth). These values are independent of the usefulness of the non-human world for human purposes.
2. Richness and diversity of life forms contribute to the realization of these values and are also values in themselves.
3. Humans have no right to reduce this richness and diversity except to satisfy vital needs.
4. The flourishing of human life and cultures is compatible with a substantially smaller human population. The flourishing of non-human life *requires* a smaller human population.
5. Present human interference with the non-human world is excessive, and the situation is rapidly worsening.
6. Policies must therefore be changed. These policies affect basic economic, technological, and ideological structures. The resulting state of affairs will be deeply different from the present.
7. The ideological change will be mainly that of appreciating life quality (dwelling in situations of inherent value) rather than adhering to an increasingly higher standard of living. There will be a profound awareness of the difference between bigness and greatness.
8. Those who subscribe to the foregoing points have an obligation directly or indirectly to try to implement the necessary changes.

Comments on the Eight Points of the Platform

RE (1): This formulation refers to the biosphere, or more professionally, to the ecosphere as a whole (this is also referred to as "ecocentrism"). This includes individuals, species, populations, habitat, as well as human and non-human cultures. Given our current knowledge of all-pervasive intimate relationships, this implies a fundamental concern and respect.

The term "life" is used here in a more comprehensive non-technical way also to refer to what biologists classify as "non-living": rivers (watersheds), landscapes, ecosystems. For supporters of deep ecology, slogans such as "let the river live" illustrate this broader usage so common in many cultures.

Inherent value, as used in (1), is common in deep ecology literature (e.g., "The presence of inherent value in a natural object is independent of any awareness, interest, or appreciation of it by any conscious being").[5]

RE (2): The so-called simple, lower, or primitive species of plants and animals contribute essentially to the richness and diversity of life. They have value in themselves and are not merely steps toward the so-called higher or rational life forms. The second principle presupposes that life itself, as a process over evolutionary time, implies an increase of diversity and richness.

Complexity, as referred to here, is different from complication. For example, urban life may be more complicated than life in a natural setting without being more complex in the sense of multifaceted quality.

RE (3): The term "vital need" is deliberately left vague to allow for considerable latitude in judgment. Differences in climate and related factors, together with differences in the structures of societies as they now exist, need to be taken into consideration.

RE (4): People in the materially richest countries cannot be expected to reduce their excessive interference with the non-human world overnight. The stabilization and reduction of the human population will take time. Hundreds of years! Interim strategies need to be developed. But in no way does this excuse the present complacency. The extreme seriousness of our current situation must first be realized. And the longer we wait to make the necessary changes, the more drastic will be the measures needed. Until deep changes are made, substantial decreases in richness and diversity are liable to occur: the rate of extinction of species will be ten to one hundred or more times greater than in any other short period of earth history.

RE (5): This formulation is mild. For a realistic assessment, see the annual reports of the Worldwatch Institute in Washington, D.C.

The slogan of "non-interference" does not imply that humans should not modify some ecosystems, as do other species. Humans have modified the earth over their entire history and will probably continue to do so. At issue is the nature and extent of such interference. The per capita destruction of wild (ancient) forests and other wild ecosystems has been excessive in rich countries; it is essential that the poor do not imitate the rich in this regard.

The fight to preserve and extend areas of wilderness and near-wilderness ("free Nature") should continue. The rationale for such preservation should focus mainly on the ecological functions of these areas (one such function: large wilderness areas are required in the biosphere for the continued evolutionary speciation of plants and animals). Most of the present designated wilderness areas and game reserves are not large enough to allow for such speciation.

RE (6): Economic growth as it is conceived of and implemented today by the industrial states is incompatible with points (1) through (5). There is only a faint resemblance between ideal sustainable forms of economic growth and the present policies of industrial societies.

Present ideology tends to value things because they are scarce and because they have a commodity value. There is prestige in vast consumption and waste (to mention only several relevant factors).

Whereas "self-determination," "local community," and "think globally, act locally," will remain key terms in the ecology of human societies, nevertheless the implementation of deep changes requires increasingly global action: Action across borders.

Governments in Third World countries are mostly uninterested in Deep Ecological issues. When institutions in the industrial societies try to promote ecological measures through Third World governments, practically nothing is accomplished (e.g., with problems of desertification). Given this situation, support for global action through non-governmental international organizations becomes increasingly important. Many of these organizations are able to act globally "from grassroots to grassroots" thus avoiding negative governmental interference.

Cultural diversity today requires advanced technology, that is, techniques that advance the basic goals of each culture. So-called soft, intermediate, and alternative technologies are steps in this direction.

RE (7): Some economists criticize the term "quality of life" because it is supposedly vague. But, on closer inspection, what they consider to be vague is actually the nonquantifiable nature of the term. One cannot quantify adequately what is important for the quality of life as discussed here, and there is no need to do so.

RE (8): There is ample room for different opinions about priorities: what should he done first; what next? What is the most urgent? What is clearly necessary to be done, as opposed to what is highly desirable but not absolutely pressing? The frontier of the environmental crisis is long and varied, and there is a place for everyone.

The above formulations of the eight points may be useful to many supporters of the deep ecology movement. But some will certainly feel that they are imperfect, even misleading. If they need to formulate in a few words what is basic to deep ecology, then they will propose an alternative set of sentences. I shall of course be glad to refer to them as alternatives. There ought to be a measure of diversity in what is considered basic and common.

Why should we call the movement "the deep ecological movement"?[6] There are at least six other designations which cover most of the same issues: "Ecological Resistance," used by John Rodman in important discussions; "The New Natural Philosophy" coined by Joseph Meeker; "Eco-philosophy," used by Sigmund Kvaloy and others to emphasize (1) a highly critical assessment of the industrial growth societies from a general ecological point of view, and (2) the ecology of the human species; "Green Philosophy and Politics" (while the term "green" is often used in Europe, in the United States "green" has a misleading association with the rather "blue" Green agricultural revolution); "Sustainable Earth Ethics," as used by G. Tyler Miller; and "Ecosophy" (eco-wisdom), which is my own favorite term. Others could be mentioned as well.

And so, why use the adjective "deep"? This question will be easier to answer after the contrast is made between shallow and deep ecological concerns. "Deep ecology" is not a philosophy in any proper academic sense, nor is it institutionalized as a religion or an ideology. Rather, what happens is that various persons come together in campaigns and direct actions. They form a circle of friends supporting the same kind of lifestyle which others may think to be "simple," but which they themselves see as rich and many-sided. They agree on a vast array of political issues, although they may otherwise support different political parties. As in all social movements, slogans and rhetoric are indispensable for in-group coherence. They react together against the same threats in a predominantly nonviolent way. Perhaps the most influential participants are artists and writers who do not articulate their insights in terms of professional philosophy, expressing themselves rather in art or poetry. For these reasons, I use the term "movement" rather than "philosophy." But it is essential that fundamental attitudes and beliefs are involved as part of the motivation for action.

4. Deep Versus Shallow Ecology

A number of key terms and slogans from the environmental debate will clarify the contrast between the shallow and the deep ecology movements.[7]

A. Pollution

Shallow Approach: Technology seeks to purify the air and water and to spread pollution more evenly. Laws limit permissible pollution. Polluting industries are preferably exported to developing countries.

Deep Approach: Pollution is evaluated from a biospheric point of view, not focusing exclusively on its effects on human health, but rather on life as a whole, including the life conditions of every species and system. The shallow reaction to acid rain, for example, is to tend to avoid action by demanding more research, and the attempt to find species of trees which will tolerate high acidity, etc. The deep approach concentrates on what is going on in the total ecosystem and calls for a high priority fight against the economic conditions and the technology responsible for producing the acid rain. The long-range concerns are one hundred years, at least.

The priority is to fight the deep causes of pollution, not merely the superficial, short-range effects. The Third and Fourth World countries cannot afford to pay the total costs of the war against pollution in their regions; consequently they require the assistance of the First and Second World countries. Exporting pollution is not only a crime against humanity, it is a crime against life in general.

B. Resources

Shallow Approach: The emphasis is upon resources for humans, especially for the present generation in affluent societies. In this view, the resources of the earth belong to those who have the technology to exploit them. There is confidence that resources will not be depleted because, as they get rarer, a high market price will conserve them, and substitutes will be found through technological progress. Further, plants, animals, and natural objects are valuable only as resources for humans. If no human use is known, or seems likely ever to be found, it does not matter if they are destroyed.

Deep Approach: The concern here is with resources and habitats for all life-forms for their own sake. No natural object is conceived of solely as a resource. This leads, then, to a critical evaluation of human modes of production and consumption. The question arises: to what extent does an increase in production and consumption foster ultimate human values? To what extent does it satisfy vital needs, locally or globally? How can economic, legal, and educational institutions be changed to counteract destructive increases? How can resource use serve the quality of life rather than the economic standard of living as generally promoted by consumerism? From a deep perspective, there is an emphasis upon an ecosystem approach rather than the consideration merely of isolated life-forms or local situations. There is a long-range maximal perspective of time and place.

C. Population

Shallow Approach: The threat of (human) "overpopulation" is seen mainly as a problem for developing countries. One condones or even applauds population increases in one's own country for short-sighted economic, military, or other reasons; an increase in the number of humans is considered as valuable in itself or as economically profitable. The issue of an "optimum population" for humans is discussed without reference to the question of an "optimum population" for other life-forms. The destruction of wild habitats caused by increasing human population is accepted as an inevitable evil, and drastic decreases of wildlife forms tend to be accepted insofar as species are not driven to extinction. Further, the social relations of animals are ignored. A long-term substantial reduction of the global human population is not seen to be a desirable goal. In addition, the right is claimed to defend one's borders against "illegal aliens," regardless of what the population pressures are elsewhere.

Deep Approach: It is recognized that excessive pressures on planetary life stem from the human population explosion. The pressure stemming from the industrial societies is a major factor, and population reduction must have the highest priority in those societies.

D. Cultural Diversity and Appropriate Technology

Shallow Approach: Industrialization of the Western industrial type is held to be the goal of developing countries. The universal adoption of Western technology is held to be compatible with cultural diversity, together with the conservation of the positive elements (from a Western perspective) of present non-industrial societies. There is a low estimate of deep cultural differences in non-industrial societies which deviate significantly from contemporary Western standards.

Deep Approach: Protection of non-industrial cultures from invasion by industrial societies. The goals of the former should not be seen as promoting lifestyles similar to those in the rich countries. Deep cultural diversity is an analogue on the human level to the biological richness and diversity of life-forms. A high priority should be given to cultural anthropology in general education programs in industrial societies.

There should be limits on the impact of Western technology upon present existing non-industrial countries and the Fourth World should be defended against foreign domination. Political and economic policies should favor subcultures within industrial societies. Local, soft technologies should allow for a basic cultural assessment of any technical innovations, together with freely expressed criticism of so-called advanced technology when this has the potential to be culturally destructive.

E. Land and Sea Ethics

Shallow Approach: Landscapes, ecosystems, rivers, and other whole entities of nature are conceptually cut into fragments, thus disregarding larger units and comprehensive gestalts. These fragments are regarded as the properties and resources of individuals, organizations or states. Conservation is argued in terms of "multiple use" and "cost/benefit analysis." The social costs and long-term global ecological costs of resource extraction and use are usually not considered. Wildlife management is conceived of as conserving nature for "future generations of humans." Soil erosion or the deterioration of ground water quality, for example, is noted as a human loss, but a strong belief in future technological progress makes deep changes seem unnecessary.

Deep Approach: The earth does not belong to humans. For example, the Norwegian landscapes, rivers, flora and fauna, and the neighboring sea are not the property of Norwegians. Similarly, the oil under the North Sea or anywhere else does not belong to any state or to humanity. And the "free nature" surrounding a local community does not belong to the local community.

Humans only inhabit the lands, using resources to satisfy vital needs. And if their non-vital needs come in conflict with the vital needs of nonhumans, then humans should defer to the latter. The ecological destruction now going on will not be cured by a technological fix. Current arrogant notions in industrial (and other) societies must be resisted.

F. Education and the Scientific Enterprise

Shallow Approach: The degradation of the environment and resource depletion requires the training of more and more "experts" who can provide advice concerning how to continue combining economic growth with maintaining a healthy environment. We are likely to need an increasingly more dominating and manipulative technology to "manage the planet" when global economic growth makes further environmental degradation inevitable. The scientific enterprise must continue giving priority to the "hard sciences" (physics and chemistry). High educational standards with intense competition in the relevant "tough" areas of learning will be required.

Deep Approach: If sane ecological policies are adopted, then education should concentrate on an increased sensitivity to non-consumptive goods, and on such consumables where there is enough for all. Education should therefore counteract the excessive emphasis upon things with a price tag. There should be a shift in concentration from the "hard" to the "soft" sciences which stress the importance of the local and global cultures. The educational objective of the World Conservation Strategy ("building support for conservation") should be given a high priority, but within the deeper framework of respect for the biosphere.

In the future, there will be no shallow environmental movement if deep policies are increasingly adopted by governments, and thus no need for a special deep ecological social movement.

5. But Why a "Deep" Ecology?

The decisive difference between a shallow and a deep ecology, in practice, concerns the willingness to question, and an appreciation of the importance of questioning, every economic and political policy in public. This questioning is both "deep" and public. It asks "why" insistently and consistently, taking nothing for granted!

Deep ecology can readily admit to the practical effectiveness of homocentric arguments:

> It is essential for conservation to be seen as central to human interests and aspirations. At the same time, people—from heads of state to the members of rural communities—will most readily he brought to demand conservation if they themselves recognize the contribution of conservation to the achievement of their needs as perceived by them, and the solution of their problems, as perceived by them.[8]

There are several dangers in arguing solely from the point of view of narrow human interests. Some policies based upon successful homocentric arguments turn out to violate or unduly compromise the objectives of deeper argumentation. Further, homocentric arguments tend to weaken the motivation to fight for necessary social change, together with the willingness to serve a great cause. In addition, the complicated arguments in human-centered conservation documents such as the World Conservation Strategy go beyond the time and ability of many people to assimilate and understand. They also tend to provoke interminable technical disagreements among experts. Special interest groups with narrow short-term exploitive objectives, which run counter to saner ecological policies, often exploit these disagreements and thereby stall the debate and steps toward effective action.

When arguing from deep ecological premises, most of the complicated proposed technological fixes need not be discussed at all. The relative merits of alternative technological proposals are pointless if our vital needs have already been met. A focus on vital issues activates mental energy and strengthens motivation. On the other hand, the shallow environmental approach, by focusing almost exclusively on the technical aspects of environmental problems, tends to make the public more passive and disinterested in the more crucial non-technical, lifestyle-related, environmental issues.

Writers within the deep ecology movement try to articulate the fundamental presuppositions underlying the dominant economic approach in terms of value priorities, philosophy, and religion. In the shallow movement, questioning and argumentation comes to a halt long before this. The deep ecology movement is therefore "the ecology movement which questions deeper." A realization of the deep changes which are required, as outlined in the deep ecology eight point platform (discussed in #3 above) makes us realize the necessity of "questioning everything."

The terms "egalitarianism," "homocentrism," "anthropocentrism," and "human chauvinism" are often used to characterize points of view on the shallow–deep spectrum. But these terms usually function as slogans which are often open to misinterpretation. They can properly imply that man is in some respects only a "plain citizen" (Aldo Leopold) of the planet on a par with all other species, but they are sometimes interpreted as denying that humans have any "extraordinary" traits, or that, in situations involving vital interests, humans have no overriding obligations towards their own kind. But this would be a mistake: they have!

In any social movement, rhetoric has an essential function in keeping members fighting together under the same banner. Rhetorical formulations also serve to provoke interest among outsiders. Of the many excellent slogans, one might mention "nature knows best," "small is beautiful," and "all things hang together." But sometimes one may safely say that nature does not always know best, that small is sometimes dreadful, and that fortunately things hang together sometimes only loosely, or not at all.

Only a minority of deep ecology supporters are academic philosophers, such as myself. And while deep ecology cannot be a finished philosophical system, this does not mean that its philosophers should not try to be as clear as possible. So a discussion of deep ecology as a derivational system may be of value to clarify the many important premise/conclusion relations.

6. Deep Ecology Illustrated as a Derivational System

Underlying the eight tenets or principles presented in section 3, there are even more basic positions and norms which reside in philosophical systems and in various world religions. Schematically we may represent the total views logically implied in the deep ecology movement by streams of derivations from the most fundamental norms and descriptive assumptions (level 1) to the particular decisions in actual life situations (level 4).

The pyramidal model has some features in common with hypothetico-deductive systems. The main difference, however, is that some sentences at the top (= deepest) level are normative, and preferably are expressed by imperatives. This makes it possible to arrive at imperatives at the lowest derivational level: the crucial level in terms of decisions. Thus, there are

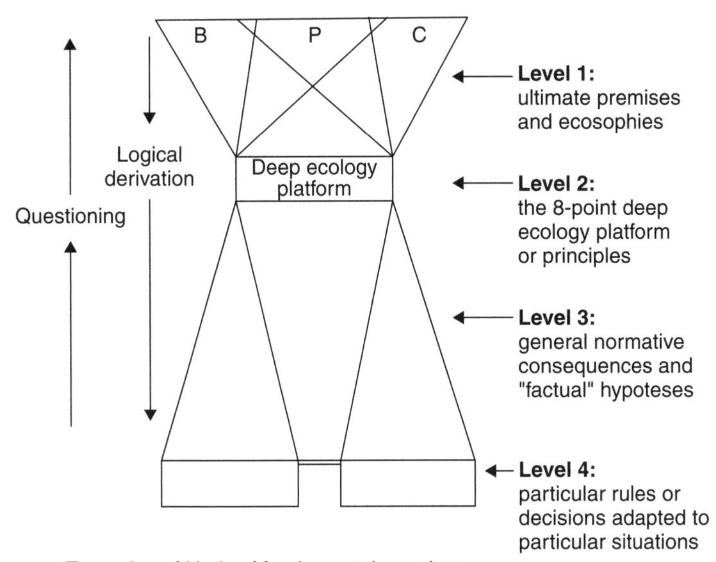

Examples of kinds of fundamental premises:
B – Buddhist
C – Christian
P – Philosophical (e.g. Spinozist or Whiteheadian)

"oughts" in our premises as well as in our conclusions. We never move from an "is" to an "ought," or vice versa. From a logical standpoint, this is decisive!

The above premise/conclusion structure (or diagram) of a total view must not be taken too seriously. It is not meant in any restrictive way to characterize creative thinking within the deep ecology movement. Creative thinking moves freely in any direction. But many of us with a professional background in science and analytical philosophy find such a diagram helpful.

As we dig deeper into the premises of our thinking, we eventually stop. Those premises we stop at are our ultimates. When we philosophize, we all stop at different places. But we all use premises which, for us, are ultimate. They belong to level 1 in the diagram. Some will use a sentence like "Every life form has intrinsic value" as an ultimate premise, and therefore place it at level 1. Others try, as I do, to conceive of it as a conclusion based on a set of premises. For these people, this sentence does not belong to level 1. There will be different ecosophies corresponding to such differences.

Obviously, point 6 of the 8 point deep ecology tenets (see section 3) cannot belong to level of the diagram. The statement "there must be new policies affecting basic economic structures" needs to be justified. If no logical justification is forthcoming, why not just assert instead that ecologically destructive "business as usual" economic policies should continue? In the diagram I have had ecosophies as ultimate premises in mind at level 1. None of the 8 points of the deep ecology principles belong at the ultimate level; they are derived as conclusions from premises at level 1.

Different supporters of the deep ecology movement may have different ultimates (level 1), but will nevertheless agree about level 2 (the 8 points). Level 4 will comprise concrete decisions in concrete situations which appear as conclusions from deliberations involving premises at levels 1 to 3. An important point: supporters of the deep ecology movement act from deep premises. They are motivated, in part, from a philosophical or religious position.

7. Multiple Roots of the Deep Ecology Platform

The deep ecology movement seriously questions the presuppositions of shallow argumentation. Even what counts as a rational decision is challenged, because what is "rational" is always defined in relation to specific aims and goals. If a decision is rational in relation to the lower level aims and goals of our pyramid, but not in relation to the highest level, then this decision should not be judged to be rational. This is an important point! If an environmentally oriented policy decision is not linked to intrinsic values or ultimates, then its rationality has yet to be determined. The deep ecology movement connects rationality with a set of philosophical or religious foundations. But one cannot expect the ultimate premises to constitute rational conclusions. There are no "deeper" premises available.

Deep ecological questioning thus reveals the fundamental normative orientations of differing positions. Shallow argumentation stops before reaching fundamentals, or it jumps from the ultimate to the particular; that is, from level 1 to level 4.

But it is not only normative claims that are at issue. Most (perhaps all) norms presuppose ideas about how the world functions. Typically the vast majority of assertions needed in normative systems are descriptive (or factual). This holds at all the levels.

As mentioned before, it does not follow that supporters of deep ecology must have identical beliefs about ultimate issues. They do have common attitudes about intrinsic values in nature, but these can, in turn (at a still deeper level), be derived from different, mutually incompatible sets of ultimate beliefs.

Thus, while a specific decision may be judged as rational from within the derivational system (if there is such) of shallow ecology, it might be judged as irrational from within the derivational system of deep ecology. Again, it should be emphasized that what is rational from within the deep ecology derivational pyramid does not require unanimity in ontology and fundamental ethics. Deep ecology as a conviction, with its subsequently derived practical recommendations, can follow from a number of more comprehensive world views, from differing ecosophies.

Those engaged in the deep ecology movement have so far revealed their philosophical or religious homes to be mainly in Christianity, Buddhism, Taoism, Baha'i, or in various philosophies. The top level of the derivational pyramid can, in such cases, be made up of normative and descriptive principles which belong to these religions and philosophies.

Since the late '70s, numerous Christians in Europe and America, including some theologians, have actively taken part in the deep ecology movement. Their interpretations of the Bible, and their theological positions in general, have been reformed from what was, until recently, a crude dominating anthropocentric emphasis.

There is an intimate relationship between some forms of Buddhism and the deep ecology movement. The history of Buddhist thought and practice, especially the principles of non-violence, non-injury, and reverence for life, sometimes makes it easier for Buddhists to understand and appreciate deep ecology than it is for Christians, despite a (sometimes overlooked) blessedness which Jesus recommended in peace-making. I mention Taoism chiefly because there is some basis for calling John Muir a Taoist, for instance, and Baha'i because of Lawrence Arturo.

Ecosophies are not religions in the classical sense. They are better characterized as *general* philosophies, in the sense of total views, inspired in part by the science of ecology. At level 1, a traditional religion may enter the derivational pyramid through a set of normative and descriptive assumptions which would be characteristic of contemporary interpretations (hermeneutical efforts) of that religion.

Supporters of the deep ecology movement act in contemporary conflicts on the basis of their fundamental beliefs and attitudes. This gives them a particular strength and a joyful expectation or hope for a greener future. But, naturally, few of them are actively engaged in a systematic verbal articulation of where they stand.

8. Ecosophy T as an Example of a Deep Ecological Derivational System

I call the ecosophy I feel at home with "Ecosophy T." My main purpose in announcing that I feel at home with Ecosophy T is didactic and dialectic. I hope to get others to announce their philosophy. If they say they have none, I maintain that they have, but perhaps don't know their own views, or are too modest or inhibited to proclaim what they believe. Following Socrates, I want to provoke questioning until others know where they stand on basic matters of life and death. This is done using ecological issues, and also by using Ecosophy T as a foil. But Socrates pretended in debate that he knew nothing. My posture seems to be the opposite. I may seem to know everything and to derive it magically from a small set of hypotheses about the world. But both interpretations are misleading! Socrates did not consistently claim to know nothing, nor do I in my Ecosophy T pretend to have comprehensive knowledge. Socrates claimed to know, for instance, about the fallibility of human claims to have knowledge.

Ecosophy T has only one ultimate norm: "Self-realization!" I do not use this expression in any narrow, individualistic sense. I want to give it an expanded meaning based on the distinction between a large comprehensive Self and narrow egoistic self as conceived of in certain Eastern traditions of *atman*.[9] This large comprehensive Self (with a capital "S") embraces all the life forms on the planet (and elsewhere?) together with their individual selves (jivas). If I were to express this ultimate norm in a few words, I would say: "Maximize (long-range, universal) Self-realization!" Another more colloquial way to express this ultimate norm would be to say "Live and let live!" (referring to all of the life forms and natural processes on the planet). If I had to give up the term fearing its inevitable misunderstanding, I would use the term "universal symbiosis." "Maximize Self-realization!" could, of course, be misinterpreted in the direction of colossal ego trips. But "Maximize symbiosis!" could be misinterpreted in the opposite direction of eliminating individuality in favor of collectivity.

Viewed systematically, not individually, maximum Self-realization implies maximizing the manifestations of all life. So next I derive the second term, "Maximize (long-range, universal) diversity!" A corollary is that the higher the levels of Self-realization attained by any person, the more any further increase depends upon the Self-realization of others. Increased self-identity involves increased identification with others. "Altruism" is a natural consequence of this identification.

This leads to a hypothesis concerning an inescapable increase of identification with other beings when one's own self-realization increases. As a result, we increasingly see ourselves in other beings, and others see themselves in us. In this way, the self is extended and deepened as a natural process of the realization of its potentialities in others.

By universalizing the above, we can derive the norm, "Self-realization for every being!" From the norm, "Maximize diversity!" and a hypothesis that maximum diversity implies a maximum of symbiosis, we can derive the norm "Maximize symbiosis!" Further, we work for life conditions such that there is a minimum of coercion in the lives of others. And so on![10] The eight points of the deep ecology platform are derived in a fairly simple way.

A philosophy as a world view inevitably has implications for practical situations. Like other ecosophies, Ecosophy T therefore moves on, without apologies, to the concrete questions of lifestyles. These will obviously show great variation because of differences in hypotheses about the world in which each of us lives, and in the "factual" statements about the concrete situations in which we make decisions.

I shall limit myself to a discussion of a couple of areas in which my "style" of thinking and behaving seem somewhat strange to friends and others who know a little about my philosophy.

First, I have a somewhat extreme appreciation of diversity; a positive appreciation of the existence of styles and behavior which I personally detest or find nonsensical (but which are not clearly incompatible with symbiosis); an enthusiasm for the "mere" diversity of species, or varieties within a genus of plants or animals; I support, as the head of a philosophy department, doctrinal theses completely at odds with my own inclinations, with the

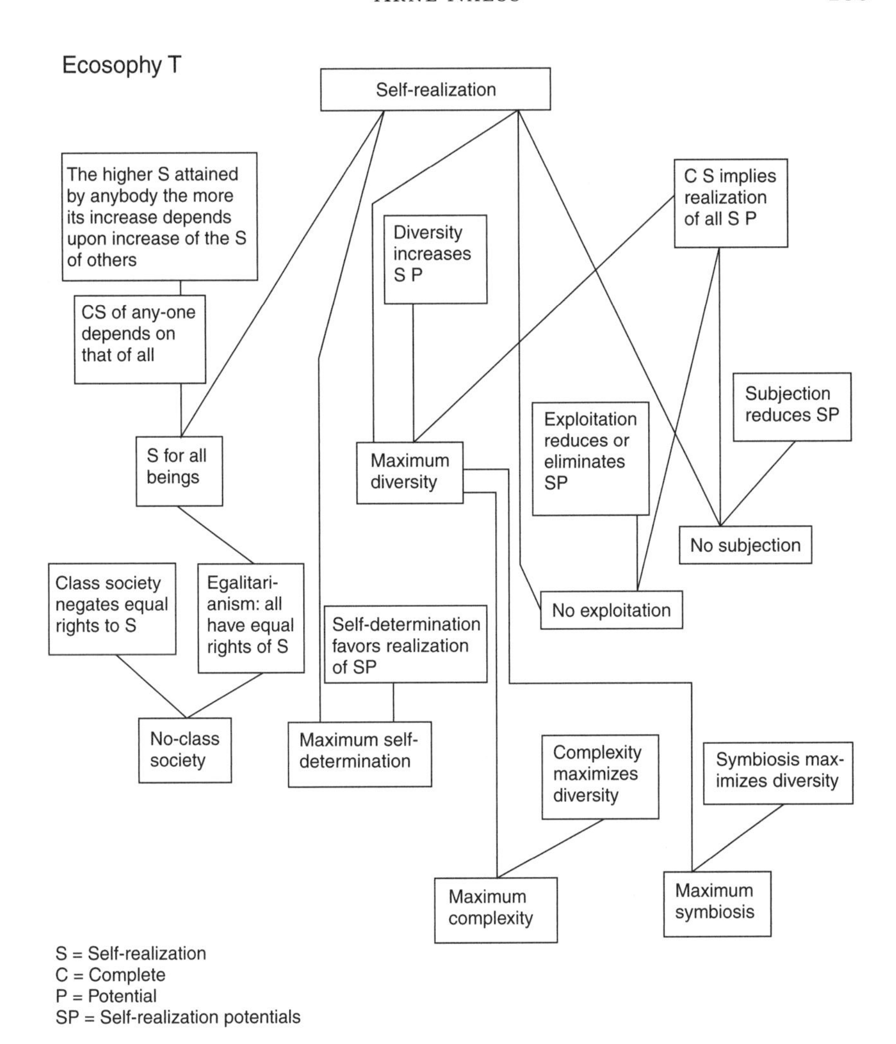

requirement only that the authors are able to understand fairly adequately some basic features of the kind of philosophy I myself feel at home with; an appreciation of combinations of seemingly incompatible interests and behaviors, which makes for an increase of subcultures within industrial states and which might to some extent help future cultural diversity. So much for "diversity!"

Second, I have a somewhat extreme appreciation of what Kant calls "beautiful actions" (good actions based on inclination), in contrast with actions which are performed out of a sense of duty or obligation. The choice of the formulation "Self-realization!" is in part motivated by the belief that maturity in humans can be measured along a scale from selfishness to an increased realization of Self, that is, by broadening and deepening the self, rather than being measured by degrees of dutiful altruism. I see joyful sharing and caring as a natural process of growth in humans.

Third, I believe that multifaceted high-level Self-realization is more easily reached through a lifestyle which is "simple in means but rich in ends" rather than through the material standard of living of the average citizens of industrial states.

The simple formulations of the deep ecology platform and Ecosophy T are not meant primarily to be used among philosophers, but also in dialogues with the "experts." When I wrote to the "experts" and environmental professionals personally, asking whether they accept the eight points of the platform, many answered positively in relation to most or all of the points. And this includes top people in the ministries of oil and energy! Nearly all were willing to let their written answers be widely published. It is an open question, however, as to what extent they will try to influence their colleagues who use only shallow argumentation. But the main conclusion to be drawn is moderately encouraging: there are views of the human/nature relationship, widely accepted among established experts responsible for environmental decisions, which require a pervasive, substantial change of present policies in favor of our "living" planet, and these views are held not only on the basis of shortsighted human interests.

NOTES

1. For more about interspecific community relationships, see Arne Naess, "Self-realization in Mixed Communities of Humans, Bears, Sheep, and Wolves," *Inquiry* 22 (1979): 321–41; Naess and Ivar Mysterud, "Philosophy of Wolf Policies I: General Principles and Preliminary Exploration of Selected Norms," *Conservation Biology* 1, 1 (1987): 22–34.
2. These problems are discussed further in Naess's keynote address to the second international Conference on Conservation Biology held at the University of Michigan in May 1985 published as "Intrinsic Value: Will the Defenders of Nature Please Rise?" *Conservation Biology* (1986): 504–15.
3. IUCN, *World Conservation Strategy: Living Resource Conservation for Sustainable Development* (Gland, Switzerland, 1980) section 13 ("Building Support for Conservation").
4. The deep ecology principles (or platform) were agreed upon during a camping trip in Death Valley, California (April, 1984) and first published in George Sessions (ed.), *Ecophilosophy* VI newsletter (May, 1984). They have subsequently appeared in a number of publications.
5. Tom Regan, "The Nature and Possibility of an Environmental Ethics," *Environmental Ethics* 3 (1981): 19–34, citation on p. 30.
6. I proposed the name "Deep, Long-Range Ecology Movement" in a lecture at the Third World Future Research conference in Bucharest in September 1972. A summary of that lecture ("The Shallow and the Deep, Long-Range Ecology Movement") was published in *Inquiry* 16 (1973): 95–100. Within the deep ecology movement it is fairly common to use the term "deep ecologist," whereas "shallow ecologist," I am glad to say, is rather uncommon. Both terms may be considered arrogant and slightly misleading. I prefer to use the awkward, but more egalitarian expression supporter of the "deep (or shallow) ecology movement," avoiding personification. Also, it is common to call deep ecology consistently anti-anthropocentric. This has led to misconceptions: see my "A Defense of the Deep Ecology Movement," *Environmental Ethics* 5 (1983).

7. The "shallow/deep" dichotomy is rough. Richard Sylvan has proposed a much more subtle classification; see his "A Critique of Deep Ecology," *Discussion Papers in Environmental Philosophy*, RSSS, Australian National University, No. 12 (1985).
8. *World Conservation Strategy*, section 13 (concluding paragraph).
9. The term *atman* is not taken in its absolutistic senses (not as a permanent indestructible "soul"). This makes it consistent with those Buddhist denials (the *avatman doctrine*) that the *atman* is to be taken in absolutist senses. Within the Christian tradition some theologians distinguish "ego" and "true self" in ways similar to these distinctions in Eastern religions. See the ecophilosophical interpretation of the gospel of Luke in Stephen Verney's *Onto the New Age* (Glasgow: Collins 1976) pp. 33–41.
10. Many authors take some steps toward derivational structures, offering mild systematizations. The chapter "Environmental Ethics and Hope" (in G. Tyler Miller, *Living in the Environment*, 3rd ed. [Belmont: Wadsworth, 1983]) is a valuable start, but the derivational relations are unclear. The logic and semantics of simple models of normative systems are briefly discussed in my "Notes on the Methodology Of Normative Systems," *Methodology and Science* 10 (1977): 64–79. For a defense of the thesis that as soon as people assert anything at all, they assume a total view, implicitly involving an ontology, methodology, epistemology, and ethics, see my "Reflections about Total Views," *Philosophy and Phenomenological Research* 25 (1964–65): 16–29. The best and wittiest warning against taking systematizations seriously is to be found in Søren Kierkegaard, *Concluding Unscientific Postscript.*

    For criticism and defense of my fundamental norm ("Self-realization"), together with my answer, see *In Sceptical Wonder: Essays in Honor of Arne Naess* (Oslo: University Press, 1982). My main exposition of Ecosophy T was originally offered in the Norwegian work, *Okologi, samfunn og livsstil* (Oslo: University Press, 5th ed., 1976). Even there, the exposition is sketchy).

29

United Church of Christ Commission for Racial Justice

In January 1986, The Commission for Racial Justice of the United Church of Christ initiated two major research projects designed to determine if there was a relationship between the treatment, storage, and disposal of hazardous waste and the racial composition of the areas where these facilities were located. The first study concentrated on commercial hazardous waste facilities; the second on toxic waste sites. These two studies were the first in the United States to consider the issue of locating hazardous waste sites in minority communities, and the ensuing national report was instrumental in bringing the issue of "environmental racism" to the attention of environmentalists, policy makers, and the general public.

"Executive Summary" from *Toxic Wastes and Race in the United States* (1987)

Recently, there has been unprecedented national concern over the problem of hazardous wastes. This concern has been focused upon the adverse environmental and health effects of toxic chemicals and other hazardous substances emanating from operating hazardous waste treatment, storage and disposal facilities as well as thousands of abandoned waste sites. Efforts to address this issue, however, have largely ignored the specific concerns of African Americans, Hispanic Americans, Asian Americans, Pacific Islanders and Native Americans. Unfortunately, racial and ethnic Americans are far more likely to be unknowing victims of exposure to such substances.

Public policies ushered in by the Reagan Administration signaled a reduction of domestic programs to monitor the environment and protect public health. Reduction of efforts to protect public health is especially disturbing in

Reprinted with the permission of Justice and Witness Ministries, United Church of Christ, *Toxic Wastes and Race in the United States: A National Report on the Racial and Socio-Economic Characteristics of Communities with Hazardous Waste Sites.* Copyright © 1987.

light of the many citizens who unknowingly may be exposed to substances emanating from hazardous waste sites. According to a December 1986 U.S. General Accounting Office (GAO) report, the U.S. Environmental Protection Agency (EPA) "does not know if it has identified 90 percent of the potentially hazardous wastes or only 10 percent."

Issues surrounding the siting of hazardous waste facilities in racial and ethnic communities gained national prominence in 1982. The Commission for Racial Justice joined ranks with residents of predominantly Black and poor Warren County, North Carolina in opposing the establishment of a polychlorinated biphenyl (PCB) disposal landfill. This opposition culminated in a nonviolent civil disobedience campaign and more than 500 arrests. As a result of the protests in Warren County. the GAO studied the racial and socio-economic status of communities surrounding four landfills in southeastern United States. It found that Blacks comprised the majority of the population in three of the four communities studied.

Previous to the Warren County demonstrations, racial and ethnic communities had been marginally involved with issues of hazardous wastes. One reason for this can be traced to the nature of the environmental movement which has historically been white middle and upper-class in its orientation. This does not mean, however, that racial and ethnic communities do not care about the quality of their environment and its effect on their lives. Throughout the course of the Commission for Racial Justice's involvement with issues of hazardous wastes and environmental pollution, we have found numerous grassroots racial and ethnic groups actively seeking to deal with this problem in their communities.

Racial and ethnic communities have been and continue to be beset by poverty, unemployment and problems related to poor housing, education and health. These communities cannot afford the luxury of being primarily concerned about the quality of their environment when confronted by a plethora of pressing problems related to their day-to-day survival. Within this context, racial and ethnic communities become particularly vulnerable to those who advocate the siting of a hazardous waste facility as an avenue for employment and economic development. Thus, proposals that economic incentives be offered to mitigate local opposition to the establishment of new hazardous waste facilities raise disturbing social policy questions.

Having observed these developments, the United Church of Christ Commission for Racial Justice decided, in 1986, to conduct extensive research on the relationship between the location of sites containing hazardous wastes and the racial and socio-economic characteristics of persons living in close proximity to those sites. The Commission for Racial Justice employed Public Data Access, Inc., a New York-based research firm, to assist in these investigations. It was hoped that these studies would lead, for the first time, to a comprehensive national analysis of the relationship between hazardous wastes and racial and ethnic communities.

"Hazardous wastes" is the term used by the EPA to define by-products of industrial production which present particularly troublesome health and environmental problems. Newly generated hazardous wastes must be managed in an approved "facility," which is defined by the EPA as any land and structures thereon which are used for treating, storing or disposing of hazardous wastes (TSD facility). Such facilities may include landfills, surface impoundments or incinerators. A "commercial" facility is defined as any facility (public or private) which accepts hazardous wastes from a third party for a fee or other remuneration.

"Uncontrolled toxic waste sites" refer to closed and abandoned sites on the EPA's list of sites which pose a present and potential threat to human health and the environment. The problem of human exposure to uncontrolled hazardous wastes is national in its scope. By 1985, the EPA had inventoried approximately 20,000 uncontrolled sites containing hazardous wastes across the nation. The potential health problems associated with the existence of these sites is highlighted by the fact that approximately 75 percent of U.S. cities derive their water supplies, in total or in part, from groundwater.

MAJOR FINDINGS

This report presents findings from two cross-sectional studies on demographic patterns associated with (1) commercial hazardous waste facilities and (2) uncontrolled toxic waste sites. The first was an analytical study which revealed a striking relationship between the location of commercial hazardous waste facilities and race. The second was a descriptive study which documented the widespread presence of uncontrolled toxic waste sites in racial and ethnic communities throughout the United States. Among the many findings that emerged from these studies, the following are most important:

Demographic Characteristics of Communities with Commercial Hazardous Waste Facilities

- Race proved to be the most significant among variables tested in association with the location of commercial hazardous waste facilities. This represented a consistent national pattern.
- Communities with the greatest number of commercial hazardous waste facilities had the highest composition of racial and ethnic residents. In communities with two or more facilities or one of the nation's five largest landfills, the average minority percentage of the population[1] was more than three times that of communities without facilities (38 percent vs. 12 percent).
- In communities with one commercial hazardous waste facility, the average minority percentage of the population was twice the average minority percentage of the population in communities without such facilities (24 percent vs. 12 percent).
- Although socio-economic status appeared to play an important role in the location of commercial hazardous waste facilities, race still proved to be more significant. This remained true after the study controlled for urbanization and regional differences. Incomes and home values were

substantially lower when communities with commercial facilities were compared to communities in the surrounding counties without facilities.

- Three out of the five largest commercial hazardous waste landfills in the United States were located in predominantly Black[2] or Hispanic communities. These three landfills accounted for 40 percent of the total estimated commercial landfill capacity in the nation.

Demographic Characteristics of Communities with Uncontrolled Toxic Waste Sites

- Three out of every five Black and Hispanic Americans lived in communities with uncontrolled toxic waste sites.
- More than 15 million Blacks lived in communities with one or more uncontrolled toxic waste sites.
- More than 8 million Hispanics lived in communities with one or more uncontrolled toxic waste sites.
- Blacks were heavily over-represented in the populations of metropolitan areas with the largest number of uncontrolled toxic waste sites. These areas include:

Memphis, TN (173 sites)	Cleveland, OH (106 sites)
St. Louis, MO (160 sites)	Chicago, IL (103 sites)
Houston, TX (152 sites)	Atlanta, GA (94 sites)

- Los Angeles, California had more Hispanics living in communities with uncontrolled toxic waste sites than any other metropolitan area in the United States.
- Approximately half of all Asian/Pacific Islanders and American Indians lived in communities with uncontrolled toxic waste sites.
- Overall, the presence of uncontrolled toxic waste sites was highly pervasive. More than half of the total population in the United States resided in communities with uncontrolled toxic waste sites.

MAJOR CONCLUSIONS AND RECOMMENDATIONS

The findings of the analytical study on the location of commercial hazardous waste facilities suggest the existence of clear patterns which show that communities with greater minority percentages of the population are more likely to be the sites of such facilities. The possibility that these patterns resulted by chance is virtually impossible,[3] strongly suggesting that some underlying factor or factors, which are related to race, played a role in the location of commercial hazardous waste facilities. Therefore, the Commission for Racial Justice concludes that, indeed, race has been a factor in the location of commercial hazardous waste facilities in the United States.

The findings of the descriptive study on the location of uncontrolled toxic waste sites suggest an inordinate concentration of such sites in Black and Hispanic communities, particularly in urban areas. This situation reveals that the issue of race is an important factor in describing the problem of uncontrolled toxic waste sites. We, therefore, conclude that the cleanup of uncontrolled toxic waste sites in Black and Hispanic communities in the United States should be given the highest possible priority.

These findings expose a serious void in present government programs addressing racial and ethnic concerns in this area. This report, therefore, strongly urges the formation of necessary offices and task forces by federal, state and local governments to fill this void. Among the many recommendations of this report, we call special attention to the following:

- We urge the President of the United States to issue an executive order mandating federal agencies to consider the impact of current policies and regulations on racial and ethnic communities.
- We urge the formation of an Office of Hazardous Wastes and Racial and Ethnic Affairs by the U.S. Environmental Protection Agency. This office should insure that racial and ethnic concerns regarding hazardous wastes, such as the cleanup of uncontrolled sites, are adequately addressed. In addition, we urge the EPA to establish a National Advisory Council on Racial and Ethnic Concerns.
- We urge state governments to evaluate and make appropriate revisions in their criteria for the siting of new hazardous waste facilities to adequately take into account the racial and socio-economic characteristics of potential host communities.
- We urge the U.S: Conference of Mayors, the National Conference of Black Mayors and the National League of Cities to convene a national conference to address these issues from a municipal perspective.
- We urge civil rights and political organizations to gear up voter registration campaigns as a means to further empower racial and ethnic communities to effectively respond to hazardous waste issues and to place hazardous wastes in racial and ethnic communities at the top of state and national legislative agendas.
- We urge local communities to initiate education and action programs around racial and ethnic concerns regarding hazardous wastes.

We also call for a series of additional actions. Of paramount importance are further epidemiological and demographic research and the provision of information on hazardous wastes to racial and ethnic communities.

This report firmly concludes that hazardous wastes in Black, Hispanic and other racial and ethnic communities should be made a priority issue at all levels of government. This issue is not currently at the forefront of the nation's attention. Therefore, concerned citizens and policy-makers, who are cognizant of this growing national problem, must make this a priority concern.

Notes

1. In this report, "minority percentage of the population" was used as a measure of "race."

2. In this report, the terminology used to describe various racial and ethnic populations was based on categories defined by the U.S. Bureau of the Census: Blacks, Hispanics, Asian/Pacific Islanders and American Indians.
3. All of the national findings were found to be statistically significant with 99.99 percent confidence (that is, findings with a probability of less than 1 in 10,000 that they occurred by chance).

30

Edward Osborne Wilson (1929–)

After earning his bachelor's and master's degrees in biology at the University of Alabama, E.O. Wilson entered the doctoral program in biology at Harvard University, earning his Ph.D. in 1955. Wilson's early work was primarily in entomology, and his studies on how ants use pheromones for communication were enormously influential. Wilson's scientific studies have also extended to the fields of ecology, sociobiology, and evolution, arguing that it is the gene, not the individual organism that is the true focus of evolution. Wilson's scientific work has won numerous honors and awards, and like Rachel Carson he has a clear, graceful writing style that makes his work accessible to a general audience as well as to scientists. He was twice awarded the Pulitzer Prize in general nonfiction for his books On Human Nature *(1979) and* The Ants *(1991).*

Much of Wilson's recent work has concerned the issue of biodiversity. In numerous articles and books such as The Diversity of Life *(1992) Wilson argues that the protection of the earth's ecosystems and the preservation of biodiversity is one of the most pressing and least often considered of our contemporary environmental problems. As he writes in "The Environmental Ethic," our lack of ecological awareness may inadvertently lead to disastrous consequences, unless we act swiftly to preserve what remains: "We should judge every scrap of biodiversity as priceless while we learn to use it and come to understand what it means to humanity. We should not knowingly allow any species or race to go extinct."*

"The Environmental Ethic" from *The Diversity of Life* (1992)

The sixth great extinction spasm of geological time is upon us, grace of mankind. Earth has at last acquired a force that can break the crucible of

Reprinted with permission of the publisher "The Environmental Ethic," from *The Diversity of Life* by Edward O. Wilson, 343–351, Cambridge, Mass.: The Belknap Press of Harvard University Press. Copyright © 1992 by Edward O. Wilson.

biodiversity. I sensed it with special poignancy that stormy night at Fazenda Dimona, when lightning flashes revealed the rain forest cut open like a cat's eye for laboratory investigation. An undisturbed forest rarely discloses its internal anatomy with such clarity. Its edge is shielded by thick secondary growth or else, along the river bank, the canopy spills down to ground level. The nighttime vision was a dying artifact, a last glimpse of savage beauty.

A few days later I got ready to leave Fazenda Dimona: gathered my muddied clothes in a bundle, gave my imitation Swiss army knife to the cook as a farewell gift, watched an overflight of Amazonian green parrots one more time, labeled and stored my specimen vials in reinforced boxes, and packed my field notebook next to a dog-eared copy of Ed McBain's police novel *Ice*, which, because I had neglected to bring any other reading matter, was now burned into my memory.

Grinding gears announced the approach of the truck sent to take me and two of the forest workers back to Manaus. In bright sunlight we watched it cross the pastureland, a terrain strewn with fire-blackened stumps and logs, the battlefield my forest had finally lost. On the ride back I tried not to look at the bare fields. Then, abandoning my tourist Portuguese, I turned inward and daydreamed. Four splendid lines of Virgil came to mind, the only ones I ever memorized, where the Sibyl warns Aeneas of the Underworld:

> The way downward is easy from Avernus.
> Black Dis's door stands open night and day.
> But to retrace your steps to heaven's air,
> There is the trouble, there is the toil . . .

For the green prehuman earth is the mystery we were chosen to solve, a guide to the birthplace of our spirit, but it is slipping away. The way back seems harder every year. If there is danger in the human trajectory, it is not so much in the survival of our own species as in the fulfillment of the ultimate irony of organic evolution: that in the instant of achieving self-understanding through the mind of man, life has doomed its most beautiful creations. And thus humanity closes the door to its past.

The creation of that diversity came slow and hard: 3 billion years of evolution to start the profusion of animals that occupy the seas, another 350 million years to assemble the rain forests in which half or more of the species on earth now live. There was a succession of dynasties. Some species split into two or several daughter species, and their daughters split yet again to create swarms of descendants that deployed as plant feeders, carnivores, free swimmers, gliders, sprinters, and burrowers, in countless motley combinations. These ensembles then gave way by partial or total extinction to newer dynasties, and so on to form a gentle upward swell that carried biodiversity to a peak—just before the arrival of humans. Life had stalled on plateaus along the way, and on five occasions it suffered extinction spasms that took 10 million years to repair. But the thrust was upward. Today the diversity of life is greater than it was a 100 million years ago—and far greater than 500 million years before that.

Most dynasties contained a few species that expanded disproportionately to create satrapies of lesser rank. Each species and its descendants, a sliver of the whole, lived an average of hundreds of thousands to millions of years. Longevity varied according to taxonomic group. Echinoderm lineages, for example, persisted longer than those of flowering plants, and both endured longer than those of mammals.

Ninety-nine percent of all the species that ever lived are now extinct. The modern fauna and flora are composed of survivors that somehow managed to dodge and weave through all the radiations and extinctions of geological history. Many contemporary world-dominant groups, such as rats, ranid frogs, nymphalid butterflies, and plants of the aster family Compositae, attained their status not long before the Age of Man. Young or old, all living species are direct descendants of the organisms that lived 3.8 billion years ago. They are living genetic libraries, composed of nucleotide sequences, the equivalent of words and sentences, which record evolutionary events all across that immense span of time. Organisms more complex than bacteria—protists, fungi, plants, animals—contain between 1 and 10 billion nucleotide letters, more than enough in pure information to compose an equivalent of the *Encyclopaedia Britannica.* Each species is the product of mutations and recombinations too complex to be grasped by unaided intuition. It was sculpted and burnished by an astronomical number of events in natural selection, which killed off or otherwise blocked from reproduction the vast majority of its member organisms before they completed their lifespans. Viewed from the perspective of evolutionary time, all other species are our distant kin because we share a remote ancestry. We still use a common vocabulary, the nucleic-acid code, even though it has been sorted into radically different hereditary languages.

Such is the ultimate and cryptic truth of every kind of organism, large and small, every bug and weed. The flower in the crannied wall—it *is* a miracle. If not in the way Tennyson, the Victorian romantic, bespoke the portent of full knowledge (by which "I should know what God and man is"), then certainly a consequence of all we understand from modern biology. Every kind of organism has reached this moment in time by threading one needle after another, throwing up brilliant artifices to survive and reproduce against nearly impossible odds.

Organisms are all the more remarkable in combination. Pull out the flower from its crannied retreat, shake the soil from the roots into the cupped hand, magnify it for close examination. The black earth is alive with a riot of algae, fungi, nematodes, mites, springtails, enchytraeid worms, thousands of species of bacteria. The handful may be only a tiny fragment of one ecosystem, but because of the genetic codes of its residents it holds more order than can be found on the surfaces of all the planets combined. It is a sample of the living force that runs the earth—and will continue to do so with or without us.

We may think that the world has been completely explored. Almost all the mountains and rivers, it is true, have been named, the coast and geodetic surveys completed, the ocean floor mapped to the deepest trenches, the atmosphere transected and chemically analyzed. The planet is now continuously monitored from space by satellites; and, not least, Antarctica, the last virgin continent, has become a research station and expensive tourist stop. The biosphere, however, remains obscure. Even though some 1.4 million species of organisms have been discovered (in the minimal sense of having specimens collected and formal scientific names attached), the total number alive on earth is somewhere between 10 and 100 million. No one can say with confidence which of these figures is the closer. Of the species given scientific names, fewer than 10 percent have been studied at a level deeper than gross anatomy. The revolution in molecular biology and medicine was achieved with a still smaller fraction, including colon bacteria, corn, fruit flies, Norway rats, rhesus monkeys, and human beings, altogether comprising no more than a hundred species.

Enchanted by the continuous emergence of new technologies and supported by generous funding for medical research, biologists have probed deeply along a narrow sector of the front. Now it is time to expand laterally, to get on with the great Linnean enterprise and finish mapping the biosphere. The most compelling reason for the broadening of goals is that, unlike the rest of science, the study of biodiversity has a time limit. Species are disappearing at an accelerating rate through human action, primarily habitat destruction but also pollution and the introduction of exotic species into residual natural environments. I have said that a fifth or more of the species of plants and animals could vanish or be doomed to early extinction by the year 2020 unless better efforts are made to save them. This estimate comes from the known quantitative relation between the area of habitats and the diversity that habitats can sustain. These area-biodiversity curves are supported by the general but not universal principle that when certain groups of organisms are studied closely, such as snails and fishes and flowering plants, extinction is determined to be widespread. And the corollary: among plant and animal remains in archaeological deposits, we usually find extinct species and races. As the last forests are felled in forest strongholds like the Philippines and Ecuador, the decline of species will accelerate even more. In the world as a whole, extinction rates are already hundreds or thousands of times higher than before the coming of man. They cannot be balanced by new evolution in any period of time that has meaning for the human race.

Why should we care? What difference does it make if some species are extinguished, if even half of all the species on earth disappear? Let me count the ways. New sources of scientific information will be lost. Vast potential biological wealth will be destroyed. Still undeveloped medicines, crops, pharmaceuticals, timber, fibers, pulp, soil-restoring vegetation, petroleum substitutes, and other products and amenities will never come to light. It is fashionable in some quarters to wave aside the small and obscure, the bugs and weeds, forgetting that an obscure moth from Latin America saved Australia's pastureland from overgrowth by cactus, that the rosy periwinkle provided the cure for Hodgkin's disease and childhood lymphocytic leukemia, that the bark of the Pacific yew offers hope for victims of ovarian and breast cancer, that a chemical from the saliva of leeches dissolves blood clots during surgery, and so on down a roster already grown long and illustrious despite the limited research addressed to it.

In amnesiac revery it is also easy to overlook the services that ecosystems provide humanity. They enrich the soil and create the very air we breathe. Without these amenities, the remaining tenure of the human race would be nasty and brief. The life-sustaining matrix is built of green plants with legions of microorganisms and mostly small, obscure animals—in other words, weeds and bugs. Such organisms support the world with efficiency because they are so diverse, allowing them to divide labor and swarm over every square meter of the earth's surface. They run the world precisely as we would wish it to be run, because humanity evolved within living communities and our bodily functions are finely adjusted to the idiosyncratic environment already created. Mother Earth, lately called Gaia, is no more than the commonality of organisms and the physical environment they maintain with each passing moment, an environment that will destabilize and turn lethal if the organisms are disturbed too much. A near infinity of other mother planets can be envisioned, each with its own fauna and flora, all producing physical environments uncongenial to human life. To disregard the diversity of life is to risk catapulting ourselves into an alien environment. We will have become like the pilot whales that inexplicably beach themselves on New England shores.

Humanity coevolved with the rest of life on this particular planet; other worlds are not in our genes. Because scientists have yet to put names on most kinds of organisms, and because they entertain only a vague idea of how ecosystems work, it is reckless to suppose that biodiversity can be diminished indefinitely without threatening humanity itself. Field studies show that as biodiversity is reduced, so is the quality of the services provided by ecosystems. Records of stressed ecosystems also demonstrate that the descent can be unpredictably abrupt. As extinction spreads, some of the lost forms prove to be keystone species, whose disappearance brings down other species and triggers a ripple effect through the demographies of the survivors. The loss of a keystone species is like a drill accidentally striking a powerline. It causes lights to go out all over.

These services are important to human welfare. But they cannot form the whole foundation of an enduring environmental ethic. If a price can be put on something, that something can be devalued, sold, and discarded. It is also possible for some to dream that people will go on living comfortably in a biologically impoverished world. They suppose that a prosthetic environment is within the power of technology, that human life can still flourish in a completely humanized world, where medicines would all be synthesized from chemicals off the shelf, food grown from a few dozen domestic crop species, the atmosphere and climate regulated by computer-driven fusion energy, and the earth made over until it becomes a literal spaceship rather than a metaphorical one, with people reading displays and touching buttons on the bridge. Such is the terminus of the philosophy of exemptionalism: do not weep for the past, humanity is a new order of life, let species die if they block progress, scientific and technological genius will find another way. Look up and see the stars awaiting us.

But consider: human advance is determined not by reason alone but by emotions peculiar to our species, aided and tempered by reason. What makes us people and not computers is emotion. We have little grasp of our true nature, of what it is to be human and therefore where our descendants might someday wish we had directed Spaceship Earth. Our troubles, as Vercors said in *You Shall Know Them*, arise from the fact that we do not know what we are and cannot agree on what we want to be. The primary cause of this intellectual failure is ignorance of our origins. We did not arrive on this planet as aliens. Humanity is part of nature, a species that evolved among other species. The more closely we identify ourselves with the rest of life, the more quickly we will be able to discover the sources of human sensibility and acquire the knowledge on which an enduring ethic, a sense of preferred direction, can be built.

The human heritage does not go back only for the conventionally recognized 8,000 years or so of recorded history, but for at least 2 million years, to the appearance of the first "true" human beings, the earliest species composing the genus *Homo*. Across thousands of generations, the emergence of culture must have been profoundly influenced by simultaneous events in genetic evolution, especially those occurring in the anatomy and physiology of the brain. Conversely, genetic evolution must have been guided forcefully by the kinds of selection rising within culture.

Only in the last moment of human history has the delusion arisen that people can flourish apart from the rest of the living world. Preliterate societies were in intimate contact with a bewildering array of life forms. Their minds could only partly adapt to that challenge. But they struggled to understand the most relevant parts, aware that the right responses gave life and fulfillment, the wrong ones sickness, hunger, and death. The imprint of that effort cannot have been erased in a few generations of urban existence. I suggest that it is to be found among the particularities of human nature, among which are these:

- People acquire phobias, abrupt and intractable aversions, to the objects and circumstances that threaten humanity in natural environments: heights, closed spaces, open spaces, running water, wolves, spiders, snakes. They rarely form phobias to the recently invented contrivances that are far more dangerous, such as guns, knives, automobiles, and electric sockets.
- People are both repelled and fascinated by snakes, even when they have never seen one in nature. In most cultures the serpent is the dominant wild animal of mythical and religious symbolism. Manhattanites dream of them with the same frequency as Zulus. This response appears to be Darwinian in origin. Poisonous snakes have been an important cause of mortality almost everywhere, from Finland to Tasmania, Canada to Patagonia; an untutored alertness in their presence saves lives. We note a kindred response in many

primates, including Old World monkeys and chimpanzees: the animals pull back, alert others, watch closely, and follow each potentially dangerous snake until it moves away. For human beings, in a larger metaphorical sense, the mythic, transformed serpent has come to possess both constructive and destructive powers: Ashtoreth of the Canaanites, the demons Fu-Hsi and Nu-kua of the Han Chinese, Mudamma and Manasa of Hindu India, the triple-headed giant Nehebkau of the ancient Egyptians, the serpent of Genesis conferring knowledge and death, and, among the Aztecs, Cihuacoatl, goddess of childbirth and mother of the human race, the rain god Tlaloc, and Quetzalcoatl, the plumed serpent with a human head who reigned as lord of the morning and evening star. Ophidian power spills over into modern life: two serpents entwine the caduceus, first the winged staff of Mercury as messenger of the gods, then the safe-conduct pass of ambassadors and heralds, and today the universal emblem of the medical profession.

- The favored living place of most peoples is a prominence near water from which parkland can be viewed. On such heights are found the abodes of the powerful and rich, tombs of the great, temples, parliaments, and monuments commemorating tribal glory. The location is today an aesthetic choice and, by the implied freedom to settle there, a symbol of status. In ancient, more practical times the topography provided a place to retreat and a sweeping prospect from which to spot the distant approach of storms and enemy forces. Every animal species selects a habitat in which its members gain a favorable mix of security and food. For most of deep history, human beings lived in tropical and subtropical savanna in East Africa, open country sprinkled with streams and lakes, trees and copses. In similar topography modern peoples choose their residences and design their parks and gardens, if given a free choice. They simulate neither dense jungles, toward which gibbons are drawn, nor dry grasslands, preferred by hamadryas baboons. In their gardens they plant trees that resemble the acacias, sterculias, and other native trees of the African savannas. The ideal tree crown sought is consistently wider than tall, with spreading lowermost branches close enough to the ground to touch and climb, clothed with compound or needle-shaped leaves.
- Given the means and sufficient leisure, a large portion of the populace backpacks, hunts, fishes, birdwatches, and gardens. In the United States and Canada more people visit zoos and aquariums than attend all professional athletic events combined. They crowd the national parks to view natural landscapes, looking from the tops of prominences out across rugged terrain for a glimpse of tumbling water and animals living free. They travel long distances to stroll along the seashore, for reasons they can't put into words.

These are examples of what I have called *biophilia*, the connections that human beings subconsciously seek with the rest of life. To biophilia can be added the idea of wilderness, all the land and communities of plants and animals still unsullied by human occupation. Into wilderness people travel in search of new life and wonder, and from wilderness they return to the parts of the earth that have been humanized and made physically secure.

Wilderness settles peace on the soul because it needs no help; it is beyond human contrivance. Wilderness is a metaphor of unlimited opportunity, rising from the tribal memory of a time when humanity spread across the world, valley to valley, island to island, godstruck, firm in the belief that virgin land went on forever past the horizon.

I cite these common preferences of mind not as proof of an innate human nature but rather to suggest that we think more carefully and turn philosophy to the central questions of human origins in the wild environment. We do not understand ourselves yet and descend farther from heaven's air if we forget how much the natural world means to us. Signals abound that the loss of life's diversity endangers not just the body but the spirit. If that much is true, the changes occurring now will visit harm on all generations to come.

The ethical imperative should therefore be, first of all, prudence. We should judge every scrap of biodiversity as priceless while we learn to use it and come to understand what it means to humanity. We should not knowingly allow any species or race to go extinct. And let us go beyond mere salvage to begin the restoration of natural environments, in order to enlarge wild populations and stanch the hemorrhaging of biological wealth. There can be no purpose more enspiriting than to begin the age of restoration, reweaving the wondrous diversity of life that still surrounds us.

The evidence of swift environmental change calls for an ethic uncoupled from other systems of belief. Those committed by religion to believe that life was put on earth in one divine stroke will recognize that we are destroying the Creation, and those who perceive biodiversity to be the product of blind evolution will agree. Across the other great philosophical divide, it does not matter whether species have independent rights or, conversely, that moral reasoning is uniquely a human concern. Defenders of both premises seem destined to gravitate toward the same position on conservation.

The stewardship of environment is a domain on the near side of metaphysics where all reflective persons can surely find common ground. For what, in the final analysis, is morality but the command of conscience seasoned by a rational examination of consequences? And what is a fundamental precept but one that serves all generations? An enduring environmental ethic will aim to preserve not only the health and freedom of our species, but access to the world in which the human spirit was born.

31

The Kyoto Protocol (1997)

Over the last twenty years scientists and the general public have become increasingly concerned about the prospect of dramatic climate change exacerbated, or even triggered, by human activity. The amount of carbon dioxide and other greenhouse gases discharged into the earth's atmosphere increased throughout the twentieth century and the nine warmest years on record have all occurred since 1990. While the implications of the greenhouse effect remain largely unknown, concern over the issue led to international support for a treaty that would reduce the amount of greenhouse gases emitted by the world's industrialized nations. From December 1 through 11, 1997, representatives from more than 160 nations met in Kyoto, Japan, to negotiate binding limitations on greenhouse gases for the developed nations pursuant to the objectives of the Framework Convention on Climate Change of 1992. The negotiated treaty, referred to as the Kyoto Protocol, called for the signatories to reduce their overall emissions of greenhouse gases by at least 5 percent below 1990 levels in the commitment period 2008–2012. Although the United States, the world's largest producer of these gases pulled out of the treaty in 2001, the treaty was ratified on November 5, 2004 when Russian President Vladimir Putin signed the treaty, which had required ratification by at least 55 countries accounting for at least 55 percent of greenhouse gas emissions in 1990.

Kyoto Protocol to the United Nations Framework Convention on Climate Change

The Parties to this Protocol, Being Parties to the United Nations Framework Convention on Climate Change, hereinafter referred to as "the Convention," In pursuit of the ultimate objective of the Convention as stated in its Article 2, Recalling the provisions of the Convention, Being guided by Article 3 of the Convention, Pursuant to the Berlin Mandate adopted by decision 1/CP.1 of the Conference of the Parties to the Convention at its first session, Have agreed as follows:

ARTICLE 1

For the purposes of this Protocol, the definitions contained in Article 1 of the Convention shall apply. In addition:

1. "Conference of the Parties" means the Conference of the Parties to the Convention.
2. "Convention" means the United Nations Framework Convention on Climate Change, adopted in New York on 9 May 1992.
3. "Intergovernmental Panel on Climate Change" means the Intergovernmental Panel on Climate Change established in 1988 jointly by the World Meteorological Organization and the United Nations Environment Programme.
4. "Montreal Protocol" means the Montreal Protocol on Substances that Deplete the Ozone Layer, adopted in Montreal on 16 September 1987 and as subsequently adjusted and amended.
5. "Parties present and voting" means Parties present and casting an affirmative or negative vote.
6. "Party" means, unless the context otherwise indicates, a Party to this Protocol.
7. "Party included in Annex I" means a Party included in Annex I to the Convention, as may be amended, or a Party which has made a notification under Article 4, paragraph 2(g), of the Convention.

ARTICLE 2

1. Each Party included in Annex I, in achieving its quantified emission limitation and reduction commitments under Article 3, in order to promote sustainable development, shall:
   (a) Implement and/or further elaborate policies and measures in accordance with its national circumstances, such as:
      (i) Enhancement of energy efficiency in relevant sectors of the national economy;
      (ii) Protection and enhancement of sinks and reservoirs of greenhouse gases not controlled by the Montreal Protocol, taking into account its commitments under relevant international environmental agreements; promotion of sustainable forest management practices, afforestation and reforestation;
      (iii) Promotion of sustainable forms of agriculture in light of climate change considerations;
      (iv) Research on, and promotion, development and increased use of, new and renewable forms of energy, of carbon dioxide

sequestration technologies and of advanced and innovative environmentally sound technologies;

(v) Progressive reduction or phasing out of market imperfections, fiscal incentives, tax and duty exemptions and subsidies in all greenhouse gas emitting sectors that run counter to the objective of the Convention and application of market instruments;

(vi) Encouragement of appropriate reforms in relevant sectors aimed at promoting policies and measures which limit or reduce emissions of greenhouse gases not controlled by the Montreal Protocol;

(vii) Measures to limit and/or reduce emissions of greenhouse gases not controlled by the Montreal Protocol in the transport sector;

(viii) Limitation and/or reduction of methane emissions through recovery and use in waste management, as well as in the production, transport and distribution of energy;

(b) Cooperate with other such Parties to enhance the individual and combined effectiveness of their policies and measures adopted under this Article, pursuant to Article 4, paragraph 2(e)(i), of the Convention. To this end, these Parties shall take steps to share their experience and exchange information on such policies and measures, including developing ways of improving their comparability, transparency and effectiveness. The Conference of the Parties serving as the meeting of the Parties to this Protocol shall, at its first session or as soon as practicable thereafter, consider ways to facilitate such cooperation, taking into account all relevant information.

2. The Parties included in Annex I shall pursue limitation or reduction of emissions of greenhouse gases not controlled by the Montreal Protocol from aviation and marine bunker fuels, working through the International Civil Aviation Organization and the International Maritime Organization, respectively.

3. The Parties included in Annex I shall strive to implement policies and measures under this Article in such a way as to minimize adverse effects, including the adverse effects of climate change, effects on international trade, and social, environmental and economic impacts on other Parties, especially developing country Parties and in particular those identified in Article 4, paragraphs 8 and 9, of the Convention, taking into account Article 3 of the Convention. The Conference of the Parties serving as the meeting of the Parties to this Protocol may take further action, as appropriate, to promote the implementation of the provisions of this paragraph.

4. The Conference of the Parties serving as the meeting of the Parties to this Protocol, if it decides that it would be beneficial to coordinate any of the policies and measures in paragraph 1(a) above, taking into account different national circumstances and potential effects, shall consider ways and means to elaborate the coordination of such policies and measures.

ARTICLE 3

1. The Parties included in Annex I shall, individually or jointly, ensure that their aggregate anthropogenic carbon dioxide equivalent emissions of the greenhouse gases listed in Annex A do not exceed their assigned amounts, calculated pursuant to their quantified emission limitation and reduction commitments inscribed in Annex B and in accordance with the provisions of this Article, with a view to reducing their overall emissions of such gases by at least 5 per cent below 1990 levels in the commitment period 2008 to 2012.
2. Each Party included in Annex I shall, by 2005, have made demonstrable progress in achieving its commitments under this Protocol.
3. The net changes in greenhouse gas emissions by sources and removals by sinks resulting from direct human-induced land-use change and forestry activities, limited to afforestation, reforestation and deforestation since 1990, measured as verifiable changes in carbon stocks in each commitment period, shall be used to meet the commitments under this Article of each Party included in Annex I. The greenhouse gas emissions by sources and removals by sinks associated with those activities shall be reported in a transparent and verifiable manner and reviewed in accordance with Articles 7 and 8.
4. Prior to the first session of the Conference of the Parties serving as the meeting of the Parties to this Protocol, each Party included in Annex I shall provide, for consideration by the Subsidiary Body for Scientific and Technological Advice, data to establish its level of carbon stocks in 1990 and to enable an estimate to be made of its changes in carbon stocks in subsequent years. The Conference of the Parties serving as the meeting of the Parties to this Protocol shall, at its first session or as soon as practicable thereafter, decide upon modalities, rules and guidelines as to how, and which, additional human-induced activities related to changes in greenhouse gas emissions by sources and removals by sinks in the agricultural soils and the land-use change and forestry categories shall be added to, or subtracted from, the assigned amounts for Parties included in Annex I, taking into account uncertainties, transparency in reporting, verifiability, the methodological work of the Intergovernmental Panel on Climate Change, the advice provided by the Subsidiary Body for Scientific and Technological Advice in accordance with Article 5 and the decisions of the Conference of the Parties. Such a decision shall apply in the second and subsequent commitment periods. A Party may choose to apply such a decision on these additional human-induced activities for its first commitment period, provided that these activities have taken place since 1990.
5. The Parties included in Annex I undergoing the process of transition to a market economy whose base year or period was established pursuant to decision 9/CP.2 of the Conference of the Parties at its second session shall use that base year or period for the implementation of their commitments under this Article. Any other Party included in Annex I undergoing

the process of transition to a market economy which has not yet submitted its first national communication under Article 12 of the Convention may also notify the Conference of the Parties serving as the meeting of the Parties to this Protocol that it intends to use an historical base year or period other than 1990 for the implementation of its commitments under this Article. The Conference of the Parties serving as the meeting of the Parties to this Protocol shall decide on the acceptance of such notification.

6. Taking into account Article 4, paragraph 6, of the Convention, in the implementation of their commitments under this Protocol other than those under this Article, a certain degree of flexibility shall be allowed by the Conference of the Parties serving as the meeting of the Parties to this Protocol to the Parties included in Annex I undergoing the process of transition to a market economy.
7. In the first quantified emission limitation and reduction commitment period, from 2008 to 2012, the assigned amount for each Party included in Annex I shall be equal to the percentage inscribed for it in Annex B of its aggregate anthropogenic carbon dioxide equivalent emissions of the greenhouse gases listed in Annex A in 1990, or the base year or period determined in accordance with paragraph 5 above, multiplied by five. Those Parties included in Annex I for whom land-use change and forestry constituted a net source of greenhouse gas emissions in 1990 shall include in their 1990 emissions base year or period the aggregate anthropogenic carbon dioxide equivalent emissions by sources minus removals by sinks in 1990 from land-use change for the purposes of calculating their assigned amount.
8. Any Party included in Annex I may use 1995 as its base year for hydrofluorocarbons, perfluorocarbons and sulphur hexafluoride, for the purposes of the calculation referred to in paragraph 7 above.
9. Commitments for subsequent periods for Parties included in Annex I shall be established in amendments to Annex B to this Protocol, which shall be adopted in accordance with the provisions of Article 21, paragraph 7. The Conference of the Parties serving as the meeting of the Parties to this Protocol shall initiate the consideration of such commitments at least seven years before the end of the first commitment period referred to in paragraph 1 above.
10. Any emission reduction units, or any part of an assigned amount, which a Party acquires from another Party in accordance with the provisions of Article 6 or of Article 17 shall be added to the assigned amount for the acquiring Party.
11. Any emission reduction units, or any part of an assigned amount, which a Party transfers to another Party in accordance with the provisions of Article 6 or of Article 17 shall be subtracted from the assigned amount for the transferring Party.
12. Any certified emission reductions which a Party acquires from another Party in accordance with the provisions of Article 12 shall be added to the assigned amount for the acquiring Party.

13. If the emissions of a Party included in Annex I in a commitment period are less than its assigned amount under this Article, this difference shall, on request of that Party, be added to the assigned amount for that Party for subsequent commitment periods.
14. Each Party included in Annex I shall strive to implement the commitments mentioned in paragraph 1 above in such a way as to minimize adverse social, environmental and economic impacts on developing country Parties, particularly those identified in Article 4, paragraphs 8 and 9, of the Convention. In line with relevant decisions of the Conference of the Parties on the implementation of those paragraphs, the Conference of the Parties serving as the meeting of the Parties to this Protocol shall, at its first session, consider what actions are necessary to minimize the adverse effects of climate change and/or the impacts of response measures on Parties referred to in those paragraphs. Among the issues to be considered shall be the establishment of funding, insurance and transfer of technology.

* * *

32

Winona LaDuke (1959–)

Although she spent much of her childhood in Los Angeles, Winona LaDuke is a member of the Mississippi band of Anishinaabeg, and moved to the White Earth reservation in northern Minnesota after earning a bachelor's degree at Harvard University 1982. At White Earth, LaDuke founded two Native rights organizations: the White Earth Land Recovery Project, which focused on raising funds to recover tribal lands; and Honor the Earth, which pursues a number of environmental initiatives. LaDuke has won numerous awards and honors for her work as an activist, and was the Green Party's vice-presidential nominee in the 1996 election.

In addition to her work as an activist, LaDuke has established a reputation as a talented writer who has published numerous essays and books on Native American and environmental issues, as well as a novel, Last Standing Woman *(1997).* In All Our Relations: Native Struggles for Land and Life *(1999), LaDuke examines cases of racial and environmental injustice on Indian reservations and offers alternatives to environmental policies that have been driven largely by corporate and technological interests.*

"The Seventh Generation" from *All Our Relations: Native Struggles for Land and Life* (1999)

Somewhere between the teachings of western science and those of the Native community there is some agreement on the state of the world. Ecosystems are collapsing, species are going extinct, the polar icecaps are melting, and nuclear bombings and accidents have contaminated the land.

According to Harvard biologist Edward O. Wilson, 50,000 species are lost every year. Three-quarters of the world's species of birds are declining, and one-quarter of all mammalian species are endangered. Tropical

Reprinted with permission of South End Press. Copyright © 1997.

rainforests, freshwater lakes, and coral reefs are at immediate risk, and global warming and climate change will accelerate the rate of biological decline dramatically.[1]

The writing is on the wall, in bold letters. There is no easy answer, and even scientists themselves seem to recognize the necessity of finding new strategies and understandings. In an unusual gathering in late 1998, for instance, NASA scientists met with Indigenous elders to discuss global warming and to hear the elders' suggestions on possible solutions. The response the scientists received may have been only part of what they had hoped for. As one observer summarized, the elders pretty much responded, "You did it, you fix it."[2]

In the final analysis, we humans can say whatever we would like—rationalize, revise statistical observations, extend deadlines, and make accommodations for a perceived "common good." But "natural law," as Yakama fisherman and former director of the Columbia Intertribal Fishing Commission Ted Strong explains, "is a hard and strict taskmaster."[3] Dump dioxin into the river, and you will inevitably eat or drink it. Assent to acceptable levels of radioactive emissions, and sooner or later, those sensitive cells in the human body will likely respond.

The challenge at the cusp of the millennium is to transform human laws to match natural laws, not vice versa. And to correspondingly transform wasteful production and voracious consumption. America and industrial society must move from a society based on conquest to one steeped in the practice of survival.

In order to do that, we must close the circle. The linear nature of industrial production itself, in which labor and technology turn natural wealth into consumer products and wastes, must be transformed into a cyclical system. In the best scenario, natural resources must be reused or not used at all, and waste production cut to a mere trickle. Those who watch carefully—*onaanaagadawaa-bandanaawaa*—know that this will require a technological, cultural, and legal transformation.

Many Indigenous teachings consider the present a time of change. Anishinaabeg teachings recognize this time of change for the people of the Seventh Fire as both a reality and an opportunity. According to these prophecies, Anishinaabeg people retrace their steps to find what was left by the trail. There are two separate roads from which to choose, for both the Anishinaabeg and those called the "light-skinned people."

Anishinaabeg elder Eddie Benton Benai, from the Lac Courte Orielles reservation in Wisconsin, is a teacher of the Anishinaabeg Midewiwin society. He discusses the two roads as

> the road to technology and the other road to Spiritualism. They [elders] feel that the road of technology represents a continuation of headlong rush to technological development. This is the road . . . that has led to modern society, to a damaged and seared earth. . . . The [other] road represents the slower path that Traditional Native people have traveled and are now seeking again. The Earth is not scorched on this trail. The grass is still growing there.[4]

A similar teaching of the Six Nations Iroquois Confederacy recognizes the importance of future generations. "In each deliberation, we must consider the impact on the seventh generation from now," they say; that is, undertake conservative thinking, and use careful deliberation. Such consideration would have preempted thousands of decisions made by the U.S. government.

Rethinking the Constitution

Walt Bresette, an Anishinaabe man from the Red Cliff reservation in northern Wisconsin, passed to the next world in early 1999. His passing was a huge loss to the Native environmental movement. But his groundbreaking work on re-envisioning the Constitution and Native treaty rights for the benefit of all people and the earth continues. Bresette was part of the Seventh Generation movement, a movement that calls for a radical amendment to the U.S. Constitution.

The preamble to the U.S. Constitution declares its intent to be to "secure the blessings of liberty, to ourselves, and our posterity." In reality, U.S. laws have been transformed by corporate interests to cater to elite interests in society. While the U.S. Constitution makes no mention of corporations, according to anti-corporate analysts Richard Grossman and Frank Adams, "the history of Constitutional law is, as former Supreme Court Justice Felix Frankfurter said, 'the history of the impact of the modern corporation on the American scene.' " Over the course of two centuries of court decisions, corporate contracts and their rates of return have been redefined as property that should be protected under the Constitution. In this way the "common good" has been redefined as "maximum corporate production and profit."[5]

Appointed judges have handed down decision after decision increasing the privileges of corporations. Corporations have been granted the power of "eminent domain" and the right to inflict "private injury and personal damage" when pursuing "progressive improvements." Most significantly, in 1886, the Supreme Court treated private corporations as "natural person[s]" protected by the Constitution and "sheltered by the Bill of Rights and the Fourteenth Amendment."[6]

Consequently, American public policy and the legal system have largely come to reflect short-term views despite the intergenerational perspective foundational to the U.S. Constitution. At the 1995 United Nations Conference on the Status of Women in Beijing, Corrine Kumar from the Asian Women's Human Rights Campaign spoke of the legal challenges in the national and international arena of this era. "The violence of the times," she explained, "has outstripped the law."[7] We have little understanding of or protection from the combined and cumulative impact of industrialism's complicated chemical soup on our bodies, ecosystems, or future generations. Public policy is lagging far behind our ability to destroy ourselves.

The rights of the people to use and enjoy air, water, and sunlight are essential to life, liberty, and the pursuit of happiness. These basic human rights have been impaired by those who discharge toxic substances into the air, water, and land. Contaminating the commons must be recognized as a fundamental wrong in our system of laws, just as defacing private property is wrong. On that basis, the Seventh Generation Amendment to the Constitution of the United States declares,

> The right of citizens of the U.S. to enjoy and use air, water, sunlight, and other renewable resources determined by the Congress to be common property shall not be impaired, nor shall such use impair their availability for use by the future generations.[8]

Bresette's other work included transforming court decisions on treaty rights into tools to transform northern Wisconsin into a sustainable, protected region. The Supreme Court's 1983 *Voigt* decision affirmed Anishinaabeg hunting, fishing, and gathering rights in ceded land in northern Wisconsin and was initially greeted with widespread outrage by non-Indians (See Chapter Six). Since then, the broader community has come to accept these rights, and Bresette and others want to expand them in ways that would benefit Indians and non-Indians alike. "A close reading of the court ruling suggests that these harvesting rights actually set extremely high environmental standards, certainly the highest in any region of the state," Bresette argued. In other words, the *Voigt* decision can be interpreted to mean not only that Indians have the right to fish and hunt in the ceded territory, but also the right to be able to "eat those fish and deer." That means that the state "should be prohibited from allowing damage to the fish by loose environmental regulation."[9]

We must follow Bresette's example and charge ourselves with curbing the rights of corporations and special interests, transforming the legal institutions of the United States back toward the preservation of the commons, and preserving everyone's rights, not just those of the economically privileged. On a community level, we must support local self-reliance and the recovery of Indigenous systems of knowledge, jurisdiction, practice, and governance.

Native people in our own reservation communities must dialogue about change, the path ahead, the options, and how we will make a better future for our children. As the conveners of the Indigenous Environmental Statement of Principles note,

> Our traditional laws lead us to understand that economic development cannot subsist on a deteriorating resource base. The environment cannot be maintained and protected when "growth" does not account for the cost of environmental and cultural destruction.[10]

The choice between the technological and the spiritual will be based on both collective and individual decisions, both simple and complex. For just as life itself is a complex web of relationships and organisms, so is the fabric of a community and a culture that chooses its future. Either way, according to Indigenous worldviews, there is no easy fix, no technological miracle.

The challenge of transformation requires the diligence and patient work evidenced by many of the people discussed in this book. And from the everglades to the subarctic, their voices for change are increasing in volume.

There is, in many Indigenous teachings, a great optimism for the potential to make positive change. Change will come. As always, it is just a matter of who determines what that change will be.

NOTES

1. Lester R. Brown, Christopher Flavin, and Hilary French, *State of the World 1997: A Worldwatch Institute Report on Progress Toward a Sustainable Society*, New York: W.W. Norton and Co., 1997, p. 13.
2. Interview with Bob Gough, November 15, 1999.
3. Ted Strong, Panel Presentation, Lewis and Clark University, October 23, 1998.
4. Eddie Benton Benai, "Seven Fires," cited in Terrnace Nelson, ed., *Okiijida: The Warrior Society*, Okiijida, Letellier, Manitoba, Canada: ROGCO, 1998, p. 4.
5. Richard Grossman and Frank Adams, in Jerry Mander and Edward Goldsmith, eds., *The Case Against the Global Economy, and For a Turn Toward the Local*, San Francisco: Sierra Club Books, 1996, p. 376.
6. Grossman and Adams, p. 384.
7. Corinne Kumar, United Nations Conference on the Status of Women, Beijing, China, September 7, 1995.
8. Walt Bresette, et al., "Seventh Generation Amendment," Anishinaabe Niiji flyer, Bayfield, WI, March 1996.
9. Interview with Walt Bresette, July 9, 1997.
10. "Indigenous Environmental Statement of Principles," Albuquerque, NM, Native Law Institute, 1995, p. 28.

Index